Fundamentals

of

Phytochemical

Analysis

Vishnu Balamurugan

i

This page intentionally left blank

Fundamentals of Phytochemical analysis

Vishnu Balamurugan

Dr.N.G.P arts and science College

Dedication

I dedicate this book to my parents, I am grateful for their belief in me and all their motivation and support. Though my parents weren't from a scientific background they motivated and encouraged me to pursue my choice of career and cheered me to excel in the same. Their unconditional love and support are the only things that keep the spark inside me truly alive.

This page intentionally left blank

Contents

Preface

Plants are used as a source of medicine for about thousands of years; the plants are used for curing simple ailments like common cold to devastating diseases like cancer. Ever thought how do these plants have such wonderful healing properties? These plants are miracles because these plants contain miraculous compounds called phytochemicals. These phytochemicals are the primary and secondary metabolites of the plants that help them survive in unfavourable conditions.

Researchers are keen on identifying and analysing these marvellous creations of nature and to find their applications in various fields of science for various applications, restoring ethno-pharmacological value of the plants. There are many books that speak about these phytochemicals and their analysing methods which trend to be admirable for many researchers they overwhelm the students with their advanced style of presentation of the information.

This book deals with the fundamentals and will be really helpful for any student enabling easy understanding of the concepts and methods for the phytochemical analysis of the plant species. As a general introduction, this book deals with an Introduction to phytochemical analysis, plant metabolites, preparation of plant samples, choosing solvents and extraction processes and qualitative and quantitative analysis of primary, secondary metabolites, vitamins, minerals and the methods that are established quite recently.

This book is written in the simple language and the fundamentals are given high emphasis for the benefit of undergraduate readers. This book can also be used as a manual for doing a research in phytochemical analysis. Understanding the fundamentals is the most important aspect in any research and I hope this book will be helpful for the readers in understanding the fundamental and most important concepts of the phytochemical analysis.

Every chapter in this book is concluded with a great reference section which will be helpful for the readers as a bibliographical guide for further advanced texts and reads.

I would like to thank my friends who has been a great support and motivated me in my life. I'm grateful for to professors of me for motivating and inspiring me with their immense knowledge. I am also thankful to the publishers for helping me out in publishing this book. I hope this book will be a great read, enjoy the ride!

Vishnu Balamurugan

June, 2019

This page intentionally left blank

Glossary

General Abbreviations

WHO- World Health Organization

UV – Ultraviolet

ATP – Adenosine tri phosphate

NADH - Nicotinamide adenine dinucleotide

NADPH - Nicotinamide adenine dinucleotide phosphate

DNA – Deoxyribonucleic acid

RNA - Ribonucleic acid

mRNA – messenger Ribonucleic acid

rRNA – ribosomal Ribonucleic acid

tRNA – transfer Ribonucleic acid

IAA - Indole 3- acetic acid

BSA – Bovine serum albumin

-OH - Hydroxyl group

TCA – Tricarboxylic acid

PLP - Pyridoxal phosphate

DTCS - N-(Dithiocarboxy) sarcosine

ALA - Acid insoluble ash

WSA - Water soluble ash

ANSA- Aminonaphtholsulfonic acid

CoA – Co-enzyme A

H –hour

L – Litre

ml – millilitre

g – gram

mg – milligram

μg – microgram

M- molarity

N- normality

kHz – kilohertz

mEq – Equivalent

R_f – retention factor

TLC – Thin layer chromatography

GC - Gas Chromatography

LC- Liquid chromatography

MS – Mass spectrometer

HPLC - High Performance Liquid Chromatography

HPTLC - High Performance Thin Layer Chromatography

OPLC - Optimum Performance Laminar Chromatography

NMR - Nuclear Magnetic Resonance

FT-ICR - Fourier transform ion cyclotron resonance

HMDB - Human Metabolome Database

Chemicals

CO_2 – carbon di oxide

HCN - hydrogen cyanide

NO_2 - Nitrogen Dioxide

NO_3 - Nitrogen trioxide

NH_4- Ammonia

H_2PO_4 - Dihydrogen phosphate

SO_4 – sulphate

MoO_4 – Molybdate

H_2SO_4 – Sulphuric acid

$CuSO_4.5H_2O$ – copper sulphate

NaOH – sodium hydroxide

HCl – Hydrochloric acid

KOH – potassium hydroxide

HNO_3 – Nitic acid

NaCl – sodium chloride

NH_3OH – ammonium hydroxide

K_2SO_4 - potassium sulphate

SeO_2 - Selenium dioxide

$((NH_4)_3BO_4$ - ammonium tetraborate

$COO.NH_4).\ H_2O$ - ammonium oxalate

$KMnO_4$ - Potassium permanganate

$NaHSO_3$ - sodium bisulphite

Na_2SO_3 - sodium sulphite

KH_2PO_4 – monopotassium phosphate

This page intentionally left blank

Chapter one

Phytochemical analysis - An introduction

"Research is to see what everybody else has seen and to think what nobody else has thought"

-Albert szent

1.1 Plants and their medicinal properties

Plants are a very important source of nutrients and a very important part in the human diet. They provide us carbohydrates, protein, vitamins, cholesterol lowering compounds, antioxidants and other important sources of biologically active substances. Many nutritional values of plants have been discussed in the literature but there is very limited research in the biologically active compounds that are present in them. These biologically active compounds are called as phytochemicals. These phytochemicals are derived from every part of the plant including roots, stem, leaves, flowers, fruits, seeds etc. These phytochemicals are sometimes used as such and in some cases they form the raw materials for a variety of other medicinally important compounds.

Medicinal plants are a gift to us from the nature as they provide a number of health benefits to us. In India these medicinal plants are used for about centuries for their properties and are still used to this date. India has a variety of traditional medical systems like Ayurveda, siddha, unani and a huge class of ethnomedicine. This knowledge of medicine was disappeared due to the western culture that has been on us on the past and is reappearing again as their importance have been realized and lack of side effects are also an important aspect in these types of traditional medicine.

As per a report by World Health Organization (WHO), over 80% of the people of developing countries are relaying on the traditional medicines that are extracted from the plants for their primary health needs. Use of these traditional medicines for the preparation of modern medical preparations is indispensable and thus 'Phytomedicines' are a link between the traditional and modern medicine.

Medicinal plants are very important in health care of individuals and communities in many developing countries. Medicinal plants are believed to be much safer and are used in treatment of various ailments .The plants provide the basic nutrients needed for the growth of animals and humans like proteins, carbohydrates, fats, vitamins and oils minerals. These plant compounds are used as alternative medicine and have become popular in western culture. They are also used in everyday medicines that we take in our daily life without even knowing that these plant compounds are present, the plant are also used as nutraceutical supplements for improving nutritional intake.

Some of the common drugs Taxol, vincristine, and morphine isolated from *Taxus brevifolia*, periwinkle, yew, and opium poppy etc., the alkaloid forskolin from *Coleus forskohlii* etc.,

1.2. Phytochemistry

Phytochemistry is a branch of chemistry which deals with the study of phytochemicals and is often considered as the subfields of botany and chemistry. The subject is also called as plant chemistry and has been developed as a discipline that is closely related to organic chemistry and plant bio chemistry. In other words, it is the chemistry of plant metabolites.

The chemicals that are produced by plants are called as phytochemicals. These are produced by the plant's primary and secondary metabolism. These phytochemicals are important for the plants to thrive or thwart other plants, animals, insects and microbial pests and pathogens. They also help plants and protect them from disease and damage caused by environmental hazards like pollution, UV, stress and draught. They are used as traditional medicine and as poisons from ancient days.

Phytochemicals are not the essential nutrients they are rather than the essential nutrients because there is no proof for them to cause any possible health effects in humans are not still established. It is known that they have roles in the protection of human health. More than 4,000 phytochemicals have been catalogued and are classified by protective function, physical characteristics and chemical characteristics.

The phytochemicals are generally classified into the following types; they include carotenoids and polyphenols which include phenolic acids, stilbenes/ lignans. Which further have classifications like flavonoids are further classified into flavones, anthocyanins, isoflavones and flavanols.

The phytochemicals are majorly classified as primary and secondary metabolites. The primary metabolites are responsible for the basic development of the plant which includes the sugars, amino acids, proteins, nucleic acids, chlorophyll, etc.

Secondary metabolites are those which are needed for the survival of the plants in a harsh environment. They forms the smell, colour and taste of the plants and secondary metabolites such as flavonoids, tannins, saponins, alkaloids, steroids, phytosterols are found to have other commercial applications like they can be used as colouring agents, as drugs as flavouring agents, insecticides, pesticides, anti-bacterial and antifungal products. Moreover they can also be used to protect humans from many diseases like cancer, diabetes, cardiovascular diseases, arthritis and aging etc.

1.3. Applications of Phytochemistry

Generally speaking, the phytochemical analysis finds applications in many fields of biological sciences even they are based on biochemical and chemical basis, even in disciplines as remote from the chemical laboratory as systematics, phytogeography, ecology and palaeobotany, phytochemical methods have become important for solving certain types of problems and will undoubtedly be used here with increasing frequency in the future.

1.3.1 Plant Physiology

Phytochemicals determine the plant physiology by determining the chemical structures, the biosynthetic pathways and the action of natural growth hormones. As a result, there are five classes of plant hormones, also known as growth regulators are recognized: they include, auxins, cytokinins, gibberellins abscisic acid and ethylene. These plant growth regulators can be of diverse composition of chemicals like terpenes (gibberellic acid) or carotenoid derivate (abscisic acid). They can be classified as growth promotors - they induce cell elongation, cell division, fruiting, flowering and formation of seeds. Examples -

Auxins, Cytokinins and gibberellins or Plant Growth Inhibitors – they cause the inhibition of growth in plants or cause abscissions or induce dormancy. Example - abscisic acid. These hormones find application in plant tissue culture and horticulture sectors.

1.3.2 Plant Pathology

Phytochemical techniques are primarily important to the pathologist for the chemical characterization· of phytotoxins and of phytoalexins (products of higher plant metabolism formed in response to microbial attack). A range of different chemical structures are involved in both cases. The most familiar phytotoxins are lycomarasmin and fusaric acid, amino acid derivatives which are wilting agents in the tomato. Other toxins that have been isolated are glycopeptides, naphthaquinones or sesquiterpenoids. Some phytotoxins are chemically labile so that special precautions have to be taken during their isolation and identification.

1.3.3 Drug Discovery

Phytochemicals are a very important source of drugs, there are number of phytochemicals that have medicinal values. Thousands of secondary metabolites were isolated and used as drug such as: Digoxin which used as cardiotonic is isolated from Digitalis purpurea (purple or common foxglove), Aescin which used as anti-inflammatory, venotonic, and as anti-eadematous drug is isolated from Aesculus hippocastanum (horse chestnut). While Ajmalicine the indole alkaloids used for treatment of circulatory disorders was first isolated from Rauwolfia sepentina.

1.3.4 Plant Ecology

The interaction of the secondary metabolites in the ecosystem is of two types, plant-plant interaction and plant- animal interaction. In both cases there are analytical problems and difficulties because of the very limited amounts of biological material at the disposal of the phytochemist for instance, the analysing of an insect's different organs for the estimation of the phytochemicals that are stored in its body by feeding on a plant is often complicated and time consuming. The important compounds that are found to have been involved in these processes are alkaloids, cardiac glycosides, steroids and terpenes etc.

1.3.5 Palaeobotany

Phytochemistry is been used in the study of fossil plants and has been used to identify the partially degraded chlorophyll pigments in lignite deposits 50 million years old Identification of terpenes in fossil resins and ambers has also yielded new data of considerable phylogenetic interest. There are many researches taken place in this field from early 1970's.

1.3.6 Plant Genetics

Phytochemistry has been providing the means of identifying anthocyanin, carotene pigments in different genotypes of the plants and is similar to the biochemical effects of the genes have a simple basis and has indicated the probable pathway of pigment synthesis in these organisms and can be mapped out by phytochemical analysis.

1.3.7 Plant Systematics

It is a hybrid between chemistry and taxonomy known as chemotaxonomy or biochemical systematics. It involves the analysis of the plant secondary metabolites and uses the information for the classification of the plants.

1.2. REFERENCES

- J. B. HARBORNE Phytochemical Methods A Guide to Modern Techniques of Plant Analysis 1973 Great Britain by Fakenham Press Limited. ISBN-13: 978-0-412-23050-9.
- John T. Arnason; Rachel Mata; John T. Romeo (2013-11-11). Phytochemistry of Medicinal Plants. Springer Science & Business Media. ISBN 9781489917782.
- "Angiosperms: Division Magnoliophyta: General Features". Encyclopædia Britannica (volume 13, 15th edition). 1993. p. 609.
- Hassan, Bassam. (2012). Medicinal Plants (Importance and Uses). Pharmaceutica Analytica Acta. 03. 10.4172/2153-2435.1000e139.
- Veeresham C. Natural products derived from plants as a source of drugs. J Adv Pharm Technol Res. 2012; 3(4):200–201. doi:10.4103/2231-4040.104709
- Hasler CM, Blumberg JB. Symposium on Phytochemicals: Biochemistry and Physiology. Journal of Nutrition 1999; 129: 756S-757S.
- Gibson EL, Wardel J, Watts CJ. Fruit and Vegetable Consumption, Nutritional

Knowledge and Beliefs in Mothers and Children. Appetite 1998; 31: 205-228.

- Katiyar C, Gupta A, Kanjilal S, Katiyar S. Drug discovery from plant sources: An integrated approach. Ayu. 2012; 33(1):10–19. doi:10.4103/0974-8520.100295

- Farag, Mohamed & Salih Mohammed, Mona & Foud, Intisar & Ahmed, Wadah & Mohamed, Malik. (2015). the role of natural products in drug discovery and development. World Journal of Pharmaceutical Research.

- Cragg G. M., Newman D. J. and Snader K. M. Natural products in drug discovery and development. Journal of Natural Products, 1997; 60: 52-60.

- CARR, D. J. ed. (1972) Plant Growth Substances, 1970. 837 pp. Springer Verlag, Be

- WOOD, R. K. S., BALLJO, A. and GRANITI, A. eds. (1972) Phytotoxins in Plant Diseases. 530 pp. Academic Press, London.

Chapter two

Metabolites and primary metabolites

2.1. Introduction

A metabolite is defined as an intermediate end product of metabolism. These metabolites have various functions like, these metabolites fuel the cell by providing energy in the form of ATP, they help in structural development, help in signalling, stimulate or inhibit the effects on enzymes or these may function as a catalyst. These can also act as a defence mechanism against other pathogens. The metabolites are classified into two major types; they are primary metabolite and secondary metabolite.

The primary metabolite is responsible for the normal functioning of the cells i.e. they are directly involved in normal growth, development and reproduction of the organism. Ex. Glucose. A Secondary metabolite is not directly responsible for the processes that are necessary for the survival of the organism but are important for its other processes that would make the environment suitable for the growth and survival of the organism. These include certain antimicrobial agents and dyes which are usually resins, alkaloids and terpenes.

2.2. Primary metabolite

A primary metabolism defined as the biological reactions that are essential for the survival of a living organism. The intermediate end products of the primary metabolism are termed as primary metabolites i.e. Most of that carbon, nitrogen, and energy ends up in molecules that are common to all cells and are required for the proper functioning of cells and organisms. These molecules, e.g., lipids, proteins, nucleic acids, and carbohydrates, are called primary metabolites.

As an example, most organisms can synthesise their own amino acids that are needed for their protein synthesis. The process of photosynthesis involves the conversion of sunlight into chemical energy, with the help of chlorophyll which leads to the synthesis of sugars and starch from the Atmospheric carbon dioxide (CO_2) by using ATP and $NADPH^+$ and these carbohydrates again broken down for the production of energy and are used for different functions of the plant metabolism.

Degradation of the carbohydrates and sugars are generally by the pathways, known as glycolysis and kerb's cycle which releases energy and organic compounds similarly fats are oxidised by the process of β-oxidation of fatty acids which also provides energy.

The important primary metabolites are carbohydrates (Sugars and starch), protein, amino acids, lipids, nucleic acids.

2.2.1. Carbohydrates

Carbohydrates are the most important primary metabolites that occupy a central position in any plant metabolism. These sugars provide the energy for the functioning of plants by photosynthesis; they are also components of cell wall and are stored means of energy. The carbohydrates are classified into three main groups, monosaccharides, oligosaccharides and polysaccharides based on their molecular size. The monosaccharides are the simplest sugars (e.g. glucose, galactose, fructose), these simples sugars fuse together to form oligosaccharides and polysaccharides. The oligosaccharides have about two to ten monosaccharides (e.g sucrose, maltose, lactose) while polysaccharides have several hundreds of monosaccharides (e.g. Starch).

There are sugar alcohols also that are percent in the plants; this is due to the readily reducing aldehyde groups in monosaccharides like pentoses and hexoses some examples for sugar alcohols are mannitol, sorbitol and glycerol.

2.2.2. Amino Acids

Amino acids contain a amine group and an acid group. The amino acids are the building blocks of proteins. In plants there are both protein amino acids and non- protein amino acids. The protein amino acids are twenty and are found in plants and animal. The non – protein amino acids are generally present in the plants.

Amino acids are also involved in the biosynthesis of proteins, alkaloids, amines, cyanogenic glycosides, porphyrins, purines, pyrimidines and cytokinins.

2.2.3. Proteins

Proteins are polymers of amino acids, several hundred of amino acids fuse together to form proteins. These proteins are coded from the nucleotide sequences in the process of central dogma of life which is DNA to mRNA, mRNA to amino acids and then amino acids to proteins.

There are different levels of protein structures they include, primary, secondary, tertiary and quaternary structures. The primary structure of protein has linear chain of amino acids while a secondary structure of protein has α-helix and β-pleated sheets. These -helixes and β- pleated sheets further fuse to form three dimensional structures of proteins called as tertiary structure and thus the quaternary structure of proteins are further more complex.

2.2.4. Nucleic acids

The nucleic acids, deoxyribonucleic acid (DNA) and ribonucleic acid (RNA) are very high molecular weight polymers of nucleic acid bases i.e. Adenine, guanine , cytosine and thiamine in DNA ; Adenine, guanine , cytosine and uracil replacing thymine in RNA with phosphate backbone and a sugar molecule. The sugar molecule is ribose sugar in RNA and deoxyribose in DNA. Molecular weight of nucleic acids ranges from 20,000 to 10,000,000. The two pyrimidine bases are cytosine and thymine (Uracil in RNA) and two purines bases are adenine and guanine. Adenine always pairs with thymine with two hydrogen bonds and guanine always pairs with cytosine with three hydrogen bonds.

DNA is the chromosomal material capable of carrying genetic information from one generation to another, these are bound together in double helical structure, and the RNA is single standard and relatively less stable than DNA in nature. The DNA is present in the nucleus of the cell and extra chromosomal DNA is found in mitochondria and chloroplast.

There are three types of RNA – messenger RNA, ribosomal RNA and transfer RNA. The messenger RNA contains the code for the translation of protein with the help of ribosome and t-RNA. The mRNA is synthesised from the coding regions (exons) of the DNA molecule in the process called transcription.

2.2.5. Lipids and fatty acids

The fatty acids in the plants are in bounded form or esterified to glycerol as fats or lipids. Lipids are present in considerable amount in some fruits and seeds, which are responsible for the germination of the seeds by providing the energy. The seeds of groundnut (Peanut), sunflower, coconut and palm etc., are used for the production of oils and fats and also in cosmetic industries. Plant fats contain unsaturated fatty acids and are important dietary requirement for humans. The important classes of fats found in plants include, phospholipids, triglycerides and glycolipids.

2.3. References

- Harris, Edward D. "Biochemical Facts behind the Definition and Properties of Metabolites" (PDF). FDA. FDA. Retrieved 28 April 2017.

- "Plant Physiology | Digital Textbook Library". www.tankonyvtar.hu. Retrieved 2016-03-16.

- http://web.nchu.edu.tw/pweb/users/taiwanfir/lesson/1146.pdf

- Buchanan, Bob B.; Gruissem, Wilhelm; Jones, Russell L. (2015-07-02). Biochemistry and

Molecular Biology of Plants. John Wiley & Sons. ISBN 9781118502198.

- Hartmann, Thomas. "From waste products to ecochemicals: fifty years research of plant secondary metabolism." Phytochemistry 68.22–24 (2007): 2831–2846. Web. 31 Mar 2011.

- "Chapter 1. Phenols, Polyphenols and Tannins: An Overview". Plant Secondary Metabolites: Occurrence, Structure and Role in the Human Diet. Nov 12, 2007. doi:10.1002/9780470988558. ISBN 9780470988558.

- Wink, Michael (26 Mar 2010). "1. Introduction: Biochemistry, Physiology and Ecological Functions of Secondary Metabolites". Annual Plant Reviews Volume 40: Biochemistry of Plant Secondary Metabolism, Second Edition. pp. 1–19. doi:10.1002/9781444320503.ch1. ISBN 9781444320503.

- BAILEY, R. W. and PRIDHAM, J. B. (1962) The separation and identification of oligosaccharides. Chroma!. Rev. 4, 114-36.

- BELL, D. J. (1962) Carbohydrates. In Comparative Biochemistry, Vol. Ill, ed.

Florkin, M. and Mason, H. S., pp. 288-355. Academic Press, New York.

- PIGMAN, W. (1957) The Carbohydrates. Academic Press, New York.

- ELMORE, R. (1968) Peptides and Proteins. pp. 154. University Press. Cambridge.

- GROSSMAN, L. and MOLDAVE, K. eds. (1967-8) Nucleic Acids, Parts A and B. In Methods in Enzymology, 2 vols. Academic Press, New York.

- NEURATH, H. ed. (1963-6) The Proteins, Composition, Structure and Function, 4 vols. Academic Press, New York.

- ANDREWS, P. G. (1964) Biochem. J. 91, 222.

- ANDREWS, P. G., HOUGH, L. and STACEY, B. (1960) Nature Lond. 185, 166.

- BAILEY, R. W. and PAIN, V. (1971) Phytochem. 10, 1065.

- BRENNER, M., NIEDERWEISER, A. and PATAKI, G. (1969) Amino acids and derivatives. In Thin Layer Chromatography, ed.

Stahl, E., pp. 730-86. George Allen and Unwin, London

- BOIT, H. G. (1961) Ergebnisse der Alkaloid Chemie bis 1960. Akademie Verlag, Berlin.

- BRENNER, M., NIEDERWEISER, A. and PATAKI, G. (1969) Amino acids and derivatives. In Thin Layer Chromatography, ed. Stahl, E., pp. 730-86. George Allen and Unwin, London.

Chapter three

Secondary metabolites

Secondary metabolites

Secondary metabolites are the compounds that are not utilized in the growth and development of the plant but are required for the survival in its environment. The secondary metabolites are used in the communication of the organisms to its mutualistic and also helpful in protecting them against pathogens and other antagonistic interactions. These are also helpful in avoiding stress tolerance, also as anti-microbial agents etc.

These can help them adapt to the biotic and abiotic stresses that a plant gets exposed due to changing environmental conditions. They also have a variety of medicinal, nutraceutical and culinary applications. There are several thousands of secondary metabolites that are estimated. Some of the most important secondary metabolites are mentioned in this book

The important secondary metabolites include nitrogen compounds, phenolic compounds, and terpenoids.

22

3.1. Nitrogen compounds

The nitrogen compounds include alkaloids, amines, cyanogenic glycosides, indoles, cytokinins and chlorophyll. The total amount of element nitrogen in a plant is 2% of its dry weight which is very lesser than the amount of carbon but still there are many different nitrogen containing organic compounds in the plants. Nitrogen is one of the first compounds to be absorbed by the plant from the soil. Nitrogen is needed for the synthesis if amino acids which are important for the production of proteins and other functional organic compounds. The alkaloids are the largest group of compounds containing nitrogen in plants. Nitrogen compounds also act as the plant growth regulators and in the synthesis of purines and pyramidine.

Alkaloids

The term alkaloid means alkali – like. These alkaloids are compounds that contain one or more nitrogen atom, form a cyclic system and have a physiological effect on humans and other animals. Most of the alkaloids are toxic in nature and hence they form a huge source of medical applications. Alkaloids are the largest class of secondary metabolites containing more than 6,000 compounds being known. They are colourless, bitter tasting substances that are liquid in room temperature. The alkaloids are known as the bitterest substances in the world, even showing significant bitterness in very less concentrations. Some of the common alkaloids are caffeine, atropine, solanine etc.,

The alkaloids are classified into two broad classifications, Non-heterocyclic or atypical alkaloids, sometimes called 'protoalkaloids' or biological amines and Heterocyclic or typical alkaloids, divided into 14 groups according to their ring structure. The heterocyclic alkanes are, Pyrrolizidine, Pyrrole and pyrrolidine, Pyridine and piperidine, Tropane, Quinoline, Isoquinoline, Aporphine, Quinolizidine, Indole or benzopyrrole, Indolizidine, Imidazole or glyoxaline, Purine, Steroidal and terpenoid.

Amines

Amines are the produced when amino acids undergo decarboxylation process. There are three main classifications of amines; they are aliphatic monoamines, aliphatic polyamines and aromatic amines. The aliphatic amines are simple compounds that are volatile in nature, for example methylamine or ethylamine. They have fishy smell and attract insects in some plants. The poly amines in contrast are not volatile but have unpleasant smell. These have growth stimulating activities due to their effect on rRNA.

Cyanogenic glycosides

Some plants can release hydrogen cyanide (HCN), this is related to the presence of some compounds called as Cyanogenic glycosides, which may give out HCN on hydrolysis either via enzymatic or non-enzymatic routes. The most common cyanogenic glycosides are linamarin and lotaustralin, compounds usually found together in plants such as flax, *Linum usitatissimum*, clover, *Trifolium repens* and birdsfoot trefoil, *Lotus corniculatus*.

Indoles

Indole 3- acetic acid (IAA) is a very important plant indole, it is a naturally occurring phyto hormone and used as a growth regulator. IAA is an auxin, Auxins promote the growth and shoots and roots in the plants, but studying auxins is a very difficult aspect because they are synthesised in very low concentrations in the plant. IAA could bind to other compounds to form an equilibrium which also makes too hard to study them. The liability of IAA in *in-vitro* and *in-vivo* environments is questionable as they rapidly oxidise into inactive compounds.

Cytokinins

Cytokinins are an important class of plant hormones that regulates the growth of the plants. They are responsible for the cell division process. There are many cytokinins that are being used now but the first identified one was kinetin and then zeatin was isolated from the *Zea mays*

Chlorophyll

Chlorophyll is the green pigment present in the chloroplast of the plants essential for the catalysing of photosynthesis process. They occur in very large amount in the plants and can be easily extracted through solvents like acetone. The chlorophyll is a molecule that has a porphyrin nucleus in the centre with magnesium atoms chelated to it and has long carbon side chains.

There are many different types of chlorophyll present in the plants, the higher plants have chlorophyll a, b which is also found in ferns and mosses and algae have chlorophyll c to e. the bacteria usually have the other types of chlorophyll.

3.2. Phenolic compounds

Phenolic compounds are the secondary metabolites that have an aromatic ring with one or more hydroxyl substitutes. The phenolic compounds are water – soluble and can be found combined with the glycosides in cells. The flavonoids are the largest group of phenolic compounds which have more than thousands of structures which are been already identified. The different types of phenolic compounds in plants are flavonoids, flavanols and flavones, anthocyanins, phenols and phenolic acids, minor Flavonoids, Xanthones and Stilbenes, quinones and Phenylpropanoids.

These phenolic compounds functions are vast and some include pigments and materials that form cell walls (lignins) but functions of many have not been still identified. All phenolic compounds are aromatic and their adsorption is intense in UV regions pf spectrum.

Flavonoids

Flavonoids are the compounds derived from the parent compound flavone. There are ten classes of flavonoids and they are Anthocyanins, Leucoanthocyanidins, Flavonols, Flavones, Glycoflavones, Biflavonyls, Chalcones and Aurones, Flavanones and Isoflavones.

Flavonoids are water soluble; they are universally present in all vascular plants where some compounds are more widely distributed, where some are not. These flavonoids are present in mixtures and it is rare to find a single flavonoid compound in plants.

Flavanols and flavones

Flavanols are found as co-pigments to anthocyanins in leaves and flowers of the plants. They occur in glycosidic combinations there are many hundreds of flavanols and are common to three compounds, quercetin, kaempferol, and myricetin. These are limited in occurrence.

The flavones lack 3-hydroxyl substitution which is the only difference from flavanols. The two common flavones are apigenin and luteolin. The flavones occur as glycosides with a carbon-carbon bond. Another series of flavanols include the biflavonyl group which has two flavones bound together with carbon- carbon or carbon oxygen bond.

Anthocyanins

Anthocyanins are one of the important classes of pigments in plants. These are water soluble and are often responsible for the red, pink, scarlet, violet and blue pigmentations in plants including flowers, leaves and fruits. The anthocyanins have a single aromatic group to which a cyaniding is bond and the colour to the compound is given by the addition or removal of a hydroxyl groups or by glycosylation or by methylation. There are more than 600 compounds of anthocyanins have been identified. Only six anthocyanidins (cyanidin, delphinidin, malvidin, pelargonidin, peonidin, and petunidin) are widely distributed in nature.

The health benefits of anthocyanins include their high antioxidant property helps in controlling diabetes, helps in prevention of obesity, cardiovascular disease prevention has anti-cancerous activities and helps in neuronal improvement and visual improvement.

Phenols and Phenolic Acids

Phenols or phenolics are a class of chemical compounds that consists of a hydroxyl group (-OH) bounded to an aromatic hydrocarbon ring. The simplest compound is phenol and there are many polyphenols too based on the number of units of phenol in the molecule.

The phenols and phenolic acids are often considered together; these acids can be combined with lignin or can be present as alcohol in the leaves. They may also present as glycosides. p-hydroxybenzoic acid, protocatechuic acid, vanillic acid and syringic acid are widely spread among the flowering plants.

Free phenols are rare in plants; hydroquinone is most widely present in plants. Phenols also have a variety of medicinal values they have high antioxidant compound content, can prevent us against chronic diseases, for instance diabetes, cancers and cardiovascular diseases. They can manage oxidative stress and can be used in the control of early stages of hyperglycemia associated with type 2 diabetes.

Minor Flavonoids, Xanthones and Stilbenes

The minor flavonoids are the classes of compounds like chalcones, aurones, flavanones, dihydrochalcones and isoflavones. These compounds are sporadic or only in few plants. They are more widely distributed and it is important to know about them while doing a phytochemical analysis.

The combination if Chalcones and aurones is known as 'anthochlors' and they are a type of yellow pigment. Flavanones are isomeric with chalcones and the two classes are interconvertible *in vitro,* but it is always not as some chalcones are found without flavanones. These flavanones have taste properties. Isoflavones are isomeric to flavanones but are rare. There are three different classes of isoflavones, they are 7,4-dihydroxyisoflavone (daidzein) and 5, 7, 4-trihydroxyisoflavone (genistein) and rotenone.

Xanthones are yellow pigments of phenolic compounds which are chemically dissimilar from flavonoids almost all the known xanthones are from four families: the Guttiferae, Gentianaceae, Moraceae and Polygalaceae.

Hydroxystilbenes and chalcones are related in biological basis but Hydroxystilbenes have one less carbon in its basic structure C_6-C_2-C_6. As they are relatively found in very less plants, they are given less interest.

Quinone

Quinone are coloured and contain a basic chromophore which is a benzoquinone itself and consists of two carbonyl groups with carbon – carbon bonds. They are usually pale yellow to black in colour. Quinones can be divided into four groups: benzoquinones, naphthoquinones, anthraquinones, and isoprenoid quinones. First three groups are generally hydroxylated, with 'phenolic' properties, and in combined form with sugar as glycoside or in a colourless, sometimes dimeric, quinol form. The isoprenoid quinones are universally distributed in plants as they are involved in photosynthesis and cellular respiration.

Phenylpropanoids

Phenylpropanoids are phenolic compounds which has three-carbon chains attached to an aromatic ring. They are derives from the amino acid, phenylalanine they contain one or more C6-C3 residues. Hydroxycoumarins, phenylpropenes and lignans are included with the Phenylpropanoids.

3.3. Terpenoids

There are so many compounds under the category of terpenoids; those compounds have the common biosynthetic origin. The terpenoids are based on the isoprene molecule $CH_2=C(CH_3)-CH=CH_2$ and two or more C_5 units. Terpenoids are classified according to the carbon atoms it contains; they include essential oils, volatile mono- and sesquiterpenes (C_{10} and C_{15}) through the less volatile diterpenes (C_{20}) to the non-volatile triterpenoids and sterols (C_{30}) and carotenoid pigments (C_{40}). These are important for the metabolism, growth, functioning and ecology of the plants.

Diterpenoids and Gibberellins

These are relatively larger compounds, which are chemically heterogeneous group of compounds which have a C_{20} carbon skeleton based on four isoprene units they are rare and are limited in distribution. Three classes of diterpenoids are resin diterpenes, toxic diterpenes and the gibberellins. The resin diterpenes compounds such as abietic and agathic acids. These resins have a protective function in nature and are exuded from woods of trees or as latex in herbaceous plants.

Gibberellins are one of the important plant hormones, but these gibberellins are produced in very minute quantity in the plants and hence they are tedious to isolate. Gibberellins are diterpenoids and there are over forty types of gibberellins known to us. They are responsible for possesses like cell elongation, flower and fruit development and also in leaf and fruit senescence.

Triterpenoids and Steroids

Triterpenoids have a carbon skeleton of six isoprene units and are derived biosynthetically from the acyclic C_{30} hydrocarbon squalene. These have complex cyclic structures and most are alcohols, acids or aldehydes. They are colourless, optically active and are crystalline in nature. Triterpenoids are classified into four main groups, true triterpenes, steroids, saponins and cardiac glycosides. The triterpenes and steroids usually occur as glycosides. Triterpenes have a bitter taste. They are also found as resins and latex in plants and trees.

Sterols are based on cyclopentane perhydrophenanthrene ring system. They are commonly present in the hormones and bile acids of animals nowadays these steroids are found in plants, these steroids found in plants are called as phytosterols. Steroids occur as both free or as glucosides in plants

Saponins are glycosides of sterols and triterpenes. These compounds are been detected in several families of the plants. They are surface active agents and have similar properties like to soap they can cause foaming and can haemolyse blood cells. The glycosidic patterns of the saponins are complex and have as many as five sugar units attached to the common compartment of glucuronic acid.

The cardiac glucosides are yet another class of triterpenoids. These are also known as cardenolides; there are many known substances with different complex mixtures of the plants. Most cardiac glucosides are toxic in nature and hence they have many pharmacological activities, as their name implies, these have beneficial activities to heart.

Carotenoids

Carotenoids are extremely distributed group of yellow pigment which is lipid soluble; these are found in all kinds of plants from bacteria to higher plants. In animals, the β- carotene is a very important dietary requirement. They are produced by splitting of vitamin A. intake of carotenoids provide animals colours, like starfish, lobsters and flamingos. The functions of carotenoids include photosynthesis (as an accessory pigment) and colouring in flowers and fruits.

There are more than 300 carotenoids known, these are either simple unsaturated hydrocarbons based on lycopene or their oxygenated derivatives. Those oxygenated derivatives are known as xanthophylls.

3.4. Other related compounds:

Sulphur compounds

The other related secondary metabolites include sulphur compounds, there are some sulphur containing amino acids like methionine and cysteine and other such sulphur containing compounds in plants, most of these compounds are volatile and also have acrid taste or an obnoxious smell which indicates their presence during isolation. The most important compound is glucosinolate found in mustard and organic disulphides that are present in *Allium* and hence the flavour of onion, garlic, mustard and radish are pungent and lastly acetylenic thiophenes found in roots and leaves of members of the *Compositae*. About seventy glucosinolates are known, the majority being aliphatic derivatives the remaining have benzyl substitutes. They have antibacterial and insecticidal properties but their applications are still under study.

Sulphides are present in *Allium* and hence can be recognized by the pungent smell and lachrymatory properties. They are found as sulphur containing amino acids but during isolation and analysis they would decompose into alkyl thiols.

The third and last class of sulphur containing compounds are thiophenes which almost occur only in a family *Compositae*. They are normally isolated along with the polyacetylenes and are purified and identified by very similar procedures.

3.5. References

* CLARKE, E. G. C. (1970) The forensic chemistry of alkaloids. In The Alkaloids, ed. Manske, H. F., Vol. XII, pp. 514-90. Academic Press, New York.

* CONN, E. E. and BUTLER, G. W. (1969) Biosynthesis of cyanogenic glycosides and other simple nitrogen compounds. In Perspectives in Phytochemistry, ed. Harborne, J. B. and Swain, T., pp. 47-74. Academic Press, London.

- Fox, J. E. (1969) The cytokinins. In Physiology of Plant Growth and Development, ed. Wilkins, M. B., pp. 85-126. McGraw-Hill, London.

- HEGNAUER, R. (1967) Comparative phytochemistry of alkaloids. In Comparative Phytochemistry, ed. Swain, T., pp. 211-30. Academic Press, London.

- HENRY, T. A. (1949) The Plant Alkaloids. Churchill, London.

- HOLDEN, M. (1965) Chlorophylls. In Chemistry and Biochemistry of Plant Pigments, ed. Goodwin, T. W., pp. 462-88. Academic Press, London.

- IKAN, R. (1969) Natural Products, A.lab. guide, pp. 178-260, Academic Press, London.

- JONES, D. A. (1972) Cyanogenic glycosides and their function. In Phytochemical Ecology, ed. Harborne, J. B., pp. 103-24. Academic Press, London.

- VERNON, L. P. and SEELEY, G. R. (1966) The Chlorophylls. Academic Press, New York.

- Tariq Pervaiz, Jiu Songtao, Faezeh Faghihi, Muhammad Salman Haider, Jinggui Fang

(2017) Naturally Occurring Anthocyanin, Structure, Functions and Biosynthetic Pathway in Fruit Plants. J Plant Biochem Physiol 5: 187. doi:10.4172/2329-9029.1000187

- Ramos, Patricio & Herrera, RaÃ°l & Moya-LeÃ³n, MarÃa. (2014). Handbook of Anthocyanins. Anthocyanins: Food sources and benefits to consumerÂ´s health.

- Derong Lin et.al An Overview of Plant Phenolic Compounds and Their Importance in Human Nutrition and Management of Type 2 Diabetes, molecules 2016, 21, 1374; doi:10.3390/molecules21101374

- GEISSMAN, T. A. ed. (1962) Chemistry of the Flavonoid Compounds. Pergamon Press, Oxford.

- HARBORNE, J. B. (1960) The chromatography of the flavonoid pigments. Chromatog. Rev. 2, 105-26.

- HARBORNE, J. B. ed. (1964) Biochemistry of Phenolic Compounds. Academic Press, London.

- HARBORNE, J. B. (1967a) Comparative Biochemistry of the Flavonoids. Academic Press, London.

- HARBORNE, J. B. (1967b) Chromatography of phenolic compounds. In Chromatography, ed. Heftmann, E., 2nd edn. pp. 677-98. Reinhold Pub. Co., New York.

- MABRY, T. J., MARKHAM, K. R., and THOMAS, M. B. (1970) The Systematic Identification of Flavonoids. Springer-Verlag, Berlin.

- MORTON, R. A. ed. (1965) Biochemistry of Quinones. Academic Press, London.

- PRIDHAM, J. B. ed. (1964) Methods in Polyphenol Chemistry. Pergamon Press Oxford.

- RIBEREAU-GAYON, P. (1972) Plant Phenolics. Oliver and Boyd, Edinburgh.

- SEIKEL, M. K. and HILLIS, W. E. (1970) Phytochem. 9, 1115.

- THOMSON, R. H. (1971) Naturally Occurring Quinones. Academic Press, London.

- BATE-SMITH, E. C. and SWAIN, T. (1966) The asperulosides and the aucubins. In Comparative Phytochemistry, ed. Swain, T., pp. 159-74. Academic Press, London.

- BASU, N. and RASTOGI, R. P. (1967) Triterpenoid saponins and sapogenins. Phytochem.6, 1249-70.

- BOLLIGER, H. R. and KOENIG, A. (1969) Carotenoids and chlorophylls. In Thin Layer Chromatography, ed. Stahl, E., pp. 266-72. George Allen and Unwin, London.

- BUSH, I. E. (1961) The Chromatography of Steroids. Pergamon Press, Oxford.

- CONNOLLY, J. D., OVERTON, K. H. and POLONSKY, J. (1970) The chemistry and biochemistry of the limonoids and quassinoids. Prog. Phytochem. 2, 385-456.

- DAVIES, B. H. (1965) Analysis of carotenoid pigments. In The Chemistry and Biochemistry of Plant Pigments, ed. Goodwin, T. W., pp. 489-532. Academic Press, London.

- GOODWIN, T. W. (1952) The Comparative Biochemistry of the Carotenoids. Chapman and Hall, London.

- GOODWIN, T. W., ed. (1965) The Chemistry and Biochemistry of Plant Pigments. Academic Press, London.

- GOODWIN, T. W. ed. (1971) Aspects of Terpenoid Chemistry and Biochemistry. Academic Press, London.

- KULSHRESHTHA, M. J., KULSHRESHTHA, D. K. and RASTOGI, R. P. (1972) The triterpenoids, Phytochem. 11,2369-81.

- NEWMAN, A. A. ed. (1972) Chemistry of Terpenes and Terpenoids. Academic Press, London.

- PRIDHAM, J. B. ed. (1967) Terpenoids in Plants. Academic Press, London.

- SIMONSEN, J. L. (1947-52) The Terpenes, 3 volumes. Cambridge Univ. Press

- APLIN, R. T., CAMBIE, R. C. and RUTLEDGE, P. S. (1963) Phytochem. 2, 205.

- ATTALAH, A. N. and NICHOLAS, H. J. (1971) Phytochem. 10, 3139.

- ETTLINGER, M. G. and KJAER, A. (1968) Sulphur compounds in plants Recent Adv. Phytochem. 1, 59-144.

- KJAER, A. (1963) The distribution of sulphur compounds. In Chemical Plant Taxonomy, ed. Swain, T., pp. 453-73. Academic Press, London.

- KJAER, A. (1966) The distribution of sulphur compounds. In Comparative Phytochemistry, ed. Swain, T., pp. 187-94. Academic Press, London.

- ASSOCIATION OF OFFICIAL AGRICULTURAL CHEMISTS (A.O.A.C.) (1965). Methods of Analysis 10th edn. p. 399. Washington DC U.S.A.

Chapter four

Miscellaneous compounds

There are some other substances other than primary and secondary metabolites, these miscellaneous compounds are found alongside with the secondary metabolites and are similar to them. The important miscellaneous compounds are Carboxylic acids, Resins, fixed oils and fats, Gums and mucilage.

4.1. Plant acids

Plants have a unique capability of accumulating organic acids in the vacuoles of the cells. For example citrus fruits can accumulate a considerable amount of acids in them for example; lemon fruits have about 58 mg of citric acid per ml at a pH of 2-5. Many leaves also have the ability to accumulate acids and they can accumulate citric, malic and isocitric acids.

There are two types of classification for these simple acids; they fall under the category of tricarboxylic acid/ Krebs cycle acids and other acids. There are nine important tricarboxylic acids and are involved in TCA cycle, they are Citric

44

acid, Malic acid, Isocitric acid, Cis-Aconitic acid, Succinic acid, Fumaric acid, Oxalacetic acid, α-Ketoglutaric acid. The most important TCA cycle acids are citric acid and malic acid, the citric acid is the most predominant of all fruit acids and next followed by malic acid. The other tricarboxylic acid cycle acids are less common.

The other plant acids are Formic acid, Acetic acid, Monoftuoracetic acid, Oxalic acid, Tartaric acid, Malonic acid, Shikimic acid, Quinic acid and Ascorbic acid. Of these, acetic acid is considered the most important due to its impotance in biosynthesis of fatty acids. They also occur in number of essential oils, phenolic acylated glycosides etc.

These organic acids are classified according to the number of carboxylic acid groups or their functional groups. The simplest carboxylic acid is formic acid HCOOH, the higher homologues are rare. The simplest dicarboxylic acid is malonic acid. The well-known tricarboxylic acid is citric acid

These organic acids are colourless, sour liquids that are soluble in water. They have relatively low melting point. They are non-volatile and are chemically stable.

4.2. Resins

Resins are a class of diterpenoids, there are three classes of diterpenoids, resin diterpenes, toxic diterpenes and the gibberellins. The resin diterpenes compounds such as abietic and agathic acids. These resins have a protective function in nature and are exuded from woods of trees or as latex in herbaceous plants.

Resins can also be phenolic compounds, highly hydroxylated or sugar substituted flavonoids are water soluble but many flavonoids are poorly water soluble or lipophilic, as are some simpler phenolics. Such lipophilic compounds are constituents of plant phenolic resins. Internally produced phenolic resins occur only in flowering plants. More commonly, lipophilic compounds are intermixed with terpenoids in angiosperm resins, particularly those covering the surface of young organs.

4.3. fixed oils and fats

Oils are alcoholic esters of fatty acids. They are synthesised from fatty acids through enzymatic action of glycerol to form esters. These are known as triglycerides. Plant fats are similar to any other fatty oils and are liquids at normal temperature. These oils occur in resins. These fats act as the reservoir of energy for the cells. These oils are found in fruits and seeds which can help in as an energy source during the germination of the seeds

The essential oils are volatile and have nice fragrance or odour; they are commercially important and are used as spices in cooking, and as natural perfumes. Mono- and sesquiterpenes are called essential oils. Monoterpenes are more abundant components of essential oils. They are found in leafs, flowers and seeds.

The sesquiterpenes also have the same skeletal structure of monoterpenes and they may be acyclic, monocyclic or bicyclic in nature. Abscisic acid is a sesquiterpene carboxylic acid, which is one of the plant growth regulators. The have inhibitory effects on the plants and cause abscissions in leaf and fruits and also cause dormancy in seeds.

4.4. Gums and mucilages

Gums

There is a misconception that gums and mucilages are same but they aren't. Gums are not resins but can be latex containing polyterpenes. The definition of gums as per a dictionary is "any plant substance that is both sticky and elastic as well as any glue used to bond surfaces". Biochemically, gums are complex chains of complex sugars (polysaccharides). A polysaccharide is a polymer of simple sugars (monosaccharides) (see chapter- 2 Carbohydrates).

Unlike resins, gums are formed under traumatic conditions like a microbial attack or insect attack as a defensive mechanism. But it is not always the condition as some gums can be formed without trauma.

Mucilages

Mucilages are like gums that have been dissolved in water as they have several similarities between them. Though they are quite similar there are many distinguishable characteristics between gums and mucilages. Mucilages are high molecular weight water soluble complexes of polysaccharides. They are closely related to cell wall components like cellulose, galactose, xylose etc. they differ from gums and can occur in various structures of plant cells like cavities, meristematic cells, surface cells , etc. they act as food reserves, helps in lubrication, in dispersion of seeds and in germination process by providing energy.

4.5. References

• ULRICH R. (1970) Organic acids. In The Biochemistry of Fruits and their Products Vol. I ed. Hulme A. C., pp. 89-118. Academic Press London.

• JEAN H. LANGENHEIM Plant Resins Chemistry, Evolution, Ecology, and Ethnobotany Timber Press, Inc 2003

• Lincoln, D. E. 1980. Leaf resin flavonoids of Diplacus aurantiacus. Biochemical Systematics and Ecology 8: 397–400.

• Lincoln, D. E. 1985. Host plant protein and phenolic resin effects on larval growth and survival of a butterfly. Journal of Chemical Ecology 11: 1459–1467.

• THOMAS, B. R. (1970) Modern and fossil plant resins. In Phytochemical Phylogeny, ed.

Harborne, J. B., pp. 59-80. Academic Press, London.

•   BASU, N. and RASTOGI, R. P. (1967) Triterpenoid saponins and sapogenins. Phytochem.6, 1249-70.

•   GUENTHER, E. (1948-52) The Essential Oils, 6 volumes. D. van Nostrand, New York.

•   Whistler, R. L. 1993. Exudate gums. In: Industrial Gums: Polysaccharides and Their Derivatives, ed. 3, pp. 309–339, eds. R. L.

•   Whistler and J. N. BeMiller. Academic Press, New York. Whistler, R. L., and J. N. BeMiller (eds.). 1993. Industrial Gums: Polysaccharides and Their Derivatives, ed. 3. Academic Press, New York.

Chapter five

Vitamins

5.1. Introduction

The term vitamin means vital amine and was coined by Casimir Funk in 1912. The vitamins are important for the human health and they have central roles in the metabolism. As vitamins are a part of our balanced dirt, it is necessary to take a lot of fresh fruits and vegetables as they have been rich sources of vitamins. As they have important metabolic functions in animals they must also have some important function in their origin too and they play an important role in plant metabolism too.

Vitamins are from a diverse group of organic molecules and they do not contain amines. Vitamins are important antioxidants, and act as enzyme cofactors, for synthesis of amino acids; they are also involved in cell wall synthesis. The deficiency of these vitamins can cause deficiency diseases in humans and hence they are very important for human metabolism too.

Though vitamins are primary metabolites, they are usually synthesised in smaller quantities, these understanding about the biosynthetic pathways of the vitamins are new and have begun an interest in researchers to understand their importance in human health and their beneficial activities to the plant.

The vitamins are classified based on their solubility into two categories; they are lipid soluble vitamins and water soluble vitamins.

5.2. Water soluble vitamins

These vitamins are soluble in water and they include Vitamin B complexes, and Vitamin C. The group of B vitamins include Vitamins B_1, B_2, B_3, B_5, B_6, B_7, B_9 and B_{12}.

Vitamin B_1

Vitamin B_1 also known as thiamine which is synthesised from the substituted pyrimidine and a thiazole coupled with a methylene bridge. It is biologically active in the form of thiamine pyrophosphate. It is essential for several enzymes in carbon central metabolism which include pyruvate dehydrogenase, transketolase, pyruvate decarboxylase and α-ketoglutarate dehydrogenase and are used throughout the cells. The deficiency of vitamin B_1 causes beri-beri in humans

Vitamin B_2

It is also called as riboflavin. They are biologically active in the form of Flavin mononucleotide (FMN). They are used ubiquitously throughout the cell. The deficiency of this disease can cause ariboflavinosis in humans.

Vitamin B_3

Vitamin B_3 is also known as niacin. The sources of niacin are nicotinic acid and nicotinamide. The biologically active form of niacin is Nicotinamide adenine dinucleotide. These are also used ubiquitously throughout the cell. The deficiency can lead to a disorder known as pellagra.

Vitamin B_5

It is also known as pantothenic acid. Pantothenic acid is formed from β-alanine and pantoic acid. It is a biologically active as 4' phosphopantetheine moiety in coenzyme A and acyl carrier protein. It functions as a cofactor for enzymes that are involved in the biosynthesis and catabolism of fatty acid synthase and pyruvate decarboxylases and in biosynthesis of some secondary metabolites. The deficiency does not cause any disease.

Vitamin B$_6$

It is also known as pyridoxine, pyridoxal and pyridoxamine. The biologically active form is Pyridoxal phosphate (PLP). They are used ion transformations of amino acids like aspartate transaminases. They are also essential in the biosynthesis of some alkaloids and other secondary metabolites. The deficiency is a risk factor for the cardiovascular diseases.

Vitamin B$_7$

Vitamin B$_7$ is also known as biotin. It is referred as Vitamin H also. The biologically active form is Biotin. It act as a cofactor for a number of enzymes they include acetyl-CoA carboxylase and pyruvate carboxylase. They are commonly present in the intestinal bacteria too and hence deficiency of biotin is rare but deficiency of biotin causes Basal ganglia disease.

Vitamin B$_9$

Vitamin B$_9$ is commonly known as folic acid. Folic acid is a compex conjugate of pteridine ring stuctures that are linked with para-aminobenzoic acid that forms pteroic acid. Folic acid is commonly available in its biologically active form as Di- and tetrahydrofolate polyglutamates. These vitamins are important for biosynthesis of several metabolic enzymes. The deficiency of this vitamin causes macrocytic anemia and in long term effects may result in DNA damage.

Vitamin B$_{12}$

Vitamil B$_{12}$ is also known as cobalamin. It is made up of complex tetrapyrrol ring structure (corrin ring) and a cobalt ion in the center. It is biologically active in the form of adenosylcobalamin and methylcobalamin. Cobalamine is not present in higher plants but are present in algae, the function of cobalamine include, cobalamin-dependent methionine synthase, methylmalonyl CoA mutase and type II ribonucleotide reductase. The deficiency of vitamil B$_{12}$ causes pernicious anaemia.

Vitamin C

The common name of Vitamin C is ascorbic acid; it is derived from glucose through the uronic acid pathway. The biologically active form of vitamin C is Ascorbic acid itself. Vitamin C is one of the essential antioxidant. It is a co-factor of many enzymes like violaxanthin de-epoxidase, prolylhydroxylase, 1- aminocyclopropane-1-carboxylate oxidase, and 9- cis-epoxycarotenoid dioxygenase. The deficiency in vitamin C causes scurvy in humans.

5.3. Lipid soluble vitamins

The lipid soluble vitamins include vitamin A, vitamin D, Vitamin E, Vitamin K and Vitamin F.

Vitamin A

The vitamin A is commonly referred as retinol. It has two vitamers, they are retinol and dehydroretinol. The biologically active component of the vitamin A are retinal which is retinaldehyde and retinoic acid. Thhe vitamin A is derived from carotenes which are a member of secondary metabolites known as carotenoids. Vitamin A can lower the risks of cancer as they are very powerful antioxidants in nature. The deficiency of vitamin A can cause nyctalopia (night blindness) progressive deficiency can lead to a condition called xerophthalmia (keratinization of cornea).

Vitamin D

Vitamin D is also known as calciferol. It is a steroidal hormone that regulates specific gene expressions in cellular responses. The biologically active form of vitamin D is calcitriol. It regulates calcium and phosphorous homeostasis. It is derived from ergosterol. The deficiency of vitamin D can cause osteomalacia in adults and rickets in children.

Vitamin E

It is also called as tocopherol. Vitamin E is a mixture containing compounds of tocopherols. Tocotrienols also have vitamin E activity; the tocopherols and tocotrienols are fat soluble vitamins and have high antioxidant properties. They are mostly found in vegetable oils like in olive oil, corn oil, soybean and sunflower oil. There are no major diseases associated with the deficiency of vitamin E

Vitamin K

These are commonly known as naphthoquinone. Vitamin K naturally occurs in plants in the form of phylloquinone. These are helpful in maintenance of clotting factors and prevent clotting of blood in human. The regular intake of vitamin K in dosage of 4 to 5 mg/ day can reduce the risks of heart attacks and strokes by preventing the formation of smaller clots in the blood vessels. The deficiency of this vitamin is rare and deficiency may cause haemorrhages.

5.4. Miscellaneous vitamin like compounds

Similarly like vitamins these compounds are also classified into two types, they are lipid soluble substances and water soluble substances. The lipid soluble substances include ubiquinone. The water soluble substances include vitamin B_4, vitamin B_8, vitamin B_{11}, vitamin B_{13}, vitamin B_{15}, vitamin U, vitamin N. some of the vitamin like compounds are,

Myo-inositol

Inositol is a compound that is closely similar to that of glucose in structure. Inositols can be also converted into phytic acid. There are no known significant benefits of inositol in humans but they are essential for plants. They are multifunctional and also have central roles in plant biochemistry and physiology. Myoinositol and its products have impact on plant growth and development. They are also used in plant tissue culture media as a supplement to promote growth.

Choline

Cholines are one of the important metabolite in plants as they are responsible for the synthesis of membrane phospholipids like phosphatidylcholine. These are one of the important cell membrane compounds and constitute about 40 to 60% of lipids in plant membranes. Choline also acts as a precursor of glycine betaine in some plants like spinach. They help the plants to tolerate environmental stresses like salinity, draught etc. they are also important nutrients for humans.

Para-aminobenzoic acid

These are produced by a number microorganism and plants and are important for their metabolism but they are not essential for humans. They are not considered as vitamins but are compounds that are related to vitamins. Para-aminobenzoic acid acts as a precursor for folate synthesis in plants and are synthesised in chloroplasts.

Carnitine

Carnitines are associated with the transfer of fatty acids from blood streams to muscle cells for the process of β- oxidation process. The synthesis occurs in insects and higher animals and the amount of carnitine in the plants are very less compared to their contents in animal cells. Carnitine plays a lesser role in lipid metabolism in plants than it does in animals. They are not considered as true vitamins

Lipoic acid

Lipoic acid has similar co enzyme function to that of thiamine. It is an essential co factor for a number of enzyme complexes. They became biologically active once they are covalently attached to one part of proteins. It has no significant effects in humans and their deficiency does not cause any disease. They are not considered as true vitamins

5.5. References

• Smith, Alison & T Croft, Martin & Moulin, MichaÃ«l & Webb, Michael. (2007). Plants need their vitamins too. Current opinion in plant biology. 10. 266-75. 10.1016/j.pbi.2007.04.009.

• . Funk C: The etiology of the deficiency diseases. J State Med 1912, 20:341-368.

• Ishikawa T, Dowdle J, Smirnoff N: Progress in manipulating ascorbic acid biosynthesis and accumulation in plants. Physiol Plant 2006, 126:343-355. 3. Keefe AD, Newton GL, Miller SL: A possible prebiotic synthesis of pantetheine, a precursor to coenzyme A. Nature 1995, 373:683-685.

• Nosaka K: Recent progress in understanding thiamin biosynthesis and its genetic regulation in Saccharomyces cerevisiae. Appl Microbiol Biotechnol 2006, 72:30-40.

• Settembre E, Begley TP, Ealick SE: Structural biology of enzymes of the thiamin biosynthesis pathway. Curr Opin Struct Biol 2003, 13:739-747

• Fischer M, Bacher A: Biosynthesis of vitamin B2 in plants. Physiol Plant 2006, 126:304-318.

• Webb ME, Smith AG, Abell C: Biosynthesis of pantothenate. Nat Prod Rep 2004, 21:695-721.

• Coxon KM, Chakauya E, Ottenhof HH, Whitney HM, Blundell TL, Abell C, Smith AG:

Pantothenate biosynthesis in higher plants. Biochem Soc Trans 2005, 33:743-746.

• Alban C, Job D, Douce R: Biotin metabolism in plants.. Annu Rev Plant Physiol Plant Mol Biol 2000, 51:17-47

• Drewke C, Leistner E: Biosynthesis of vitamin B6 and structurally related derivatives. In Vitamins and Hormones, Vol 61. Edited by Litwack G, Bh thiamin pyrophosphate, revealing the mode of binding.

• Warren MJ, Raux E, Schubert HL, Escalante-Semerena JC: The biosynthesis of adegley TP. Academic Press; 2001:121-155

• Croft MT, Warren MJ, Smith AG: Algae need their vitamins. Eukaryot Cell 2006, 5:1175-1183.

• Matamoros MA, Loscos J, Coronado MJ, Ramos J, Sato S, Testillano PS, Tabata S, Becana M: Biosynthesis of ascorbic acid in legume root nodules. Plant Physiol 2006, 141:1068-1077.

• Ishikawa T, Dowdle J, Smirnoff N: Progress in manipulating ascorbic acid biosynthesis and accumulation in plants. Physiol Plant 2006, 126:343-355.

• Frank A. Loewus, Pushpalatha P.N. Murthy, myo-Inositol metabolism in plants, Plant Science,Volume 150, Issue 1, 2000 1-19, ISSN 0168-9452,

• Datko88: Datko AH, Mudd SH (1988). "Phosphatidylcholine synthesis. Differing patterns in soybean and carrot." Plant Physiol. 88; 854-861.

• McNeil01: McNeil SD, Nuccio ML, Ziemak MJ, Hanson AD (2001). "Enhanced synthesis of choline and glycine betaine in transgenic tobacco plants that overexpress phosphoethanolamine N-methyltransferase." Proc Natl Acad Sci U S A 98(17);10001-5. PMID: 11481443

• Mou02: Mou Z, Wang X, Fu Z, Dai Y, Han C, Ouyang J, Bao F, Hu Y, Li J (2002). "Silencing of phosphoethanolamine N-methyltransferase results in temperature-sensitive male sterility and salt hypersensitivity in Arabidopsis." Plant Cell 14(9);2031-43. PMID: 12215503

• Eoin P. Quinlivan, Sanja Roje, Gilles Basset, Yair Shachar-Hill, Jesse F. Gregory III and Andrew D. Hanson. The Folate Precursor p-

Aminobenzoate Is Reversibly Converted to Its Glucose Ester in the Plant Cytosol. The Journal of Biological Chemistry (2003) 278, 20731-20737.

• Bourdin B, Adenier H, Perrin Y. Carnitine is associated with fatty acid metabolism in plants. Plant Physiol Biochem. 2007 Dec;45 (12):926-31. Epub 2007Sep 29. PubMed PMID: 17988884.

• Wada M, Yasuno R, Jordan SW, Cronan JE Jr, Wada H. Lipoic acid metabolism in Arabidopsis thaliana: cloning and characterization of a cDNA encoding lipoyltransferase. Plant Cell Physiol. 2001 Jun;42(6):650-6. PubMed PMID 11427685.

Chapter six

Minerals

6.1. Introduction

Minerals have a very big role in the metabolism of plants; both the high concentration and low concentration of minerals in the plants have severe effects in their metabolism. These minerals are important for the biosynthesis of certain secondary metabolites which provide them antioxidant, antibacterial, antifungal, pesticidal and insecticidal activities. The contents of minerals in the soil vary and due to which there are several adverse effects in the metabolic activities of plants. Different stages of plants need different micro and macro nutrients for their growth and development.

The minerals are classified into two types based on the requirement of the plants, they are, macro nutrients and micro nutrients. In essence, some minerals are needed for plants in large quantities and they are classified as macro nutrients. Some minerals are required in minimal quantities and hence are termed as micronutrients. The different types of the macro and micro nutrients needed for plants are as following.

6.2. Macro nutrients

The elements that are needed in higher concentration are called as macro nutrients; these are the most important minerals that the plants need for the survival. They are essential for the normal plant physiology and biochemistry. They are needed for the basic metabolism of plants. There are nine important macro nutrients they are, Carbon (C), Hydrogen (H), Oxygen (O), Nitrogen (N), Phosphorus (P), Potassium (K), Calcium (Ca), Magnesium (Mg) and Sulphur (S).

Carbon

The carbon, hydrogen and oxygen are the most important macro nutrients for the plants; these are not only absorbed from the soil but are also taken from the atmosphere. Carbon is taken up by the plants in the form of CO_2 and enters into the photosynthesis where these atoms are converted into sugars and then as starch. The carbon forms the almost all the compounds in the plants and it is responsible for the energy production in the plants. In other words they are the building blocks of all the organic compounds in the plants. Carbon is the most abundant element in the plants with 76% of total dry weight.

Hydrogen

Hydrogen is taken up by the plants from the soil in the form of water (H_2O), hydrogen is used in the process of photosynthesis and oxygen is released into the atmosphere. The hydrogen and carbon forms the basic skeleton of most of the organic compounds and hence hydrogen is one of the important elements in the plants. Plants contain about 10% of hydrogen in the dry weight.

Oxygen

Oxygen is obtained by the plants from the atmosphere as CO_2. The plants need oxygen for their survival and metabolism. Oxygen aids in the process of aerobic respiration in plants where oxygen is the final oxygen acceptor. Krebs cycle also needs Oxygen for proper functioning. Oxygen is also an important constituent in many organic compounds like carbohydrates, fats, amino acids, nucleic acids and many other organic compounds. Oxygen is about 10.5% of total dry weight of the plant.

Nitrogen

Nitrogen is one of the most important elements needed for the growth and metabolism of plants.

They are absorbed in the form of NO_2, NO_3 or NH_4 from the soil. The nitrogen is one of the important compounds in many primary and secondary metabolites they include amino acids, proteins, nucleic acids, vitamins and plant growth regulators. They are required in all parts of the plants and predominately in meristematic tissues. It is about 1-3 % of total dry weight of plant. The deficiency in nitrogen can cause abscission in leaves, reduction in growth, lateral bud dormancy, inhibition of cell division etc.

Phosphorous

They are important constituents of DNA, RNA, phospholipids and coenzymes of NAD and NADP and are also an important compound in ATP. They are obtained from soil in the form of $H_2PO_4^-$ They are found in high concentration at meristematic region i.e. growing tips of the plants. They are involved in the synthesis of amino acids, and ATP. The phospholipids are also important as they are the important component of the cell membrane. They are usually about 0.3 to 6 % of plant's dry weight. The deficiency causes retarded growth, abscissions and also interferes in the development of vascular system of the plant.

Potassium

The potassium is obtained from the soil as potassium ions (K^+). The function of potassium in plants is poorly understood similarly as nitrogen and phosphorus, potassium is also abundant in meristematic tissues, it is an activator of enzymes in several carbohydrate metabolisms. They also function in the synthesis of peptide bonds. They are about 0.05 to 1.0% of plant's dry weight. The deficiency can cause defects in photosynthesis, chlorophyll development, respiration and water retention in plants.

Calcium

They are obtained from the soil as calcium ions (Ca^{2+}). These are important compounds that have a role in the cell walls. They form calcium pectate in the cell walls. The middle lamellas of the plant cells have calcium and magnesium. The calcium is also important for the formation of cell membranes and lipid structures. They are also involved in mitosis and activate a variety of enzymes. The deficiency affects meristematic tissue and eventually dies and terminates the growth. The cell walls become brittle, vacuolization, cell enlargement and differentiation occurs in regions near shoot apex. The total dry weight is 0.1 – 3.5 % of total plant mass

Magnesium

Magnesium is one of the most important minerals as they are directly associated with photosynthesis. Chlorophyll contains magnesium without which the process of photosynthesis will not occur. They are also involved in carbohydrate metabolism and synthesis of nucleic acids. They are also associated in protein synthesis. The deficiency causes extensive interveinal chlorosis and due to which the anthocyanin pigments may appear. They are absorbed as Mg^{2+} ions from soil and their content is 0.05-0.07 % of plant mass.

Sulphur

The sulphur is obtained as SO_4^{2+} ions from soil. They participate in the protein structure as there are sulphur containing amino acids like cysteine and methionine are present. They are also found in the vitamins like thiamine and biotin and in some co-enzymes. They also form disulphide bonds in variety of compounds and most importantly in proteins which is important for stabilization of proteins. Deficiency causes chlorosis and is similar to that of magnesium deficiency may also cause accumulation of starch and sugar in cells. They are about 0.05-1.5 % of total dry weight.

6.3. Micro nutrients

Micro nutrients are the minerals and elements that are required for normal functioning and metabolism of plant but in minute quantities. There are few micro nutrients that are essential for plant growth and metabolism and they are, iron, copper, manganese, zinc, molybdenum, cobalt, boron, chlorine and sodium.

Iron

Iron is also essential for the synthesis of chlorophyll and maintenance of chloroplast, it also have a critical role in the synthesis of DNA, respiration and photosynthesis. Iron is also needed for the activation of many metabolic pathways. It is also responsible for many biochemical and physiological functions in plants. They are components of various flavoproteins and iron porphyrins. Deficiency causes chlorosis.

Copper

Copper is an essential element for the normal growth and metabolism of plants, they act as a co-factor for number of compounds including many metalloproteins. They are also found in phenolases, lactace and vitamin C. But still high amount of copper in cells can cause problems to the cells by inhabiting plant growth and cellular processes and hence plants control the copper content by haemostasis. Deficiency causes necrosis in tips of young leaves.

Manganese

Manganese is an essential micronutrient for plants; they are absorbed as divalent Mn^{2+} and are accumulated in parenchymatous cells and petioles of leaves, manganese is involves in photosynthesis, respiration and in metabolism of nitrogen. Deficiency causes inhibition of cell elongation and necrotic spots in interveins of leaves

Zinc

Zinc is a micronutrient that is absorbed as divalent Zn^{2+} ions from soil. They are important for the biosynthesis of plant hormones called as auxin, Indole acetic acid (IAA). It also serves as a activator of several enzymes and also takes part in metabolism of plants. Deficiency causes discoloration, shorter internodes and growth is stunted.

Molybdenum

Molybdenum id absorbed from the as MoO_4^{3-} , they function in nitrogen fixation in the plants, and regulates the vitamin C content in plants. They form molybdenum cofactor (Moco) by binding to organic moiety of molybdopterin. These have high redox properties. Deficiency leads to drop of vitamin C in the plants.

Cobalt

Cobalt is essential compound for the many enzymes and co-enzymes; they show effects in the plant growth and metabolism. They are absorbed as Co^{2+} ions from soil. Cobalt does not have any roles in respiratory chains.

Boron

There are many functions suggested for boron, they include cell differentiation, nitrogen, fats and hormone metabolism and also have roles in photosynthesis. Sexual reproduction of plants is sensitive to boron. They also have roles in cell wall integrity, phenol metabolisms and nitrogen fixation. Deficiency causes death of shoot tips and flowers do not from in deficiency of boron in plants.

Chlorine

Chlorines are absorbed as Cl⁻ ions from the soil. They are essential for higher plants and have many metabolic processes. They supports plant growth and regulates ion balance and in evolution of oxygen in photosynthesis. They are also helpful in tolerance of stress and disease tolerance

Sodium

The sodium is widely taken up by plants but is not considered to be essential, classifying plant nutrients into "essential nutrients" and "functional nutrients". Sodium is classified under "functional nutrients", they are required for increasing biomass of the plants and can replace potassium in many ways. And used for cell enlargement etc. though they are not essential for land plants, they are helpful in maintaining osmotic and ionic balance in marine plants.

6.4. References

- Mishra, Brijkishore & Rastogi, Dr Anu & Shukla, Sudhir. (2012). Regulatory Role of Mineral Elements in the Metabolism of Medicinal Plants.

• Ukpong I. J., Abasiekong B. O. and Etuk B. A. Phytochemical screening and mineral elements composition of Xanthosoma sagittifolium inflorescence.

Asian Journal of Plant Science and Research, 2014, 4(6): 32-35

• Coruzzi G, Bush DR (2001) Nitrogen and carbon nutrient and metabolite signaling in plants. Plant Physiology 125 (1), 61-64

• Runguphan W, Qu X, O'Connor SE (2010) Integrating carbon-halogen bond formation into medicinal plant metabolism. Nature 468, 461-464

• Crawford NM, Kahn ML, Leustek T, Long SR (2000) Nitrogen and sulfur. In: Buchanan BB, Gruissen W, Jones RL (Eds) Biochemistry and Molecular Biology of Plants, Americans Society of Plant Biologists, Rockville, MD, pp 824-849

• Demeyer K, Degaegere R (1998) Nitrogen and alkaloid accumulation and partitioning in Datura stramonium L. Journal of Herbs, Spices and Medicinal Plants 5 (3), 15-23

• Mazid M, Khan TA, Mohammad F (2011) Response of crop plants under sulphur stress tolerance: A holistic approach. Journal of Stress Physiology and Biochemistry 7 (3), 23-57

• Rout, Gyana. (2015). Role of iron in plant growth and metabolism. Reviews in Agricultural Sciences. 3. 1-2. 10.7831/ras.3.1.

• Joardar Mukhopadhyay, M., & Sharma, A. (1991). Manganese in cell metabolism of higher plants. The Botanical Review, 57(2), 117–149. https://doi.org/10.1007/BF02858767

• Yruela, Inmaculada. (2005). Copper in plants. Brazilian Journal of Plant Physiology, 17(1), 145-156. https://dx.doi.org/10.1590/S1677-04202005000100012

• JONES, W. N. (1935). Zinc and Plant Metabolism. Nature, 136(3442), 646. https://doi.org/10.1038/136646c0

• 1: Tejada-Jiménez M, Chamizo-Ampudia A, Galván A, Fernández E, Llamas Á.Molybdenum metabolism in plants. Metallomics. 2013 Sep;5(9):1191-203. doi: 10.1039/c3mt00078h. Review. PubMed PMID: 23800757.

• Palit, Syamasri & Sharma, Archana & Talukder, Geeta. (1994). Effects of Cobalt on

Plants. The Botanical Review. 60. 149-181. 10.1007/BF02856575.

• Ahmad, Waqar & Niaz, A & Kanwal, Shamsa & Khalid, M. (2009). Role of boron in plant growth: a review. Journal of agricultural research. 47.

• Chen, W., He, Z. L., Yang, X. E., Mishra, S., & Stoffella, P. J. (2010). CHLORINE NUTRITION OF HIGHER PLANTS: PROGRESS AND PERSPECTIVES. Journal of Plant Nutrition, 33(7), 943–952. https://doi.org/10.1080/01904160903242417

• Subbarao, Guntur & Ito, Osamu & L Berry, W & Wheeler, Ray & Mohammad Dr, Referee & , Pessarakli. (2003). Sodiumâ€"A Functional Plant Nutrient *. Critical Reviews in Plant Sciences. 22. 391-. 10.1080/07352680390243495.

Chapter seven

Preparation of plant sample

7.1. Introduction

Before analysis of any plant sample, they are needed to be prepared for the analysis. The preparation steps include selection of the plant sample, collection of the plant sample, identification of the samples, cleaning of the samples, drying and grinding of the samples for further analysis.

7.2. Selection of plant samples

Proper selection and identification is important for any phytochemical research any defect in this could severely affect the research and may reduce the value of the study.

1. Traditionally used plants by humans for food, medicine or poison based on literature or other sources can be investigated.
2. A random or systematic collection of plants over a large biodiverse area regarding secondary metabolite production can be used.

3. Other plant species that are phylogenetically related to the species known to produce a compound of interest.
4. Species based on the reports of biological activity in the literature.
5. Research about their distribution, abundance and reproductive biology must also be done.
6. The plant should be abundant in the specific site of collection.
7. Large and matured parts of the plants should be used
8. Consideration must be done on the yield of the desired compound while selection itself.

7.3. Collection of plant samples

Plant collection can be done on either from wild forests or from herbariums. But in the case of wild plants there is a risk of plants that are been incorrectly identified. They have an advantage that they do not contain any pesticides or herbicides. After collection they are processed soon to prevent the deterioration of secondary metabolites present in the samples.

1. Collection should not interfere in the vegetative growth of the plants and should ensure that they can produce new

vegetative growth even after collection of samples.

2. While collecting a whole plant, some plants must be left in the collection site for reproduction.

3. While collecting enough foliage must be left for the plants to continue the growth and reproduction.

4. While collecting samples from larger trees and shrubs, samples must be collected from the side branches rather than the trunk of the branches.

5. Harvesting of samples like barks and roots needs special care.

6. Same plants from different localities can be collected and they may have different levels of the active compound

7. Collection can be done in different time of days and night and in different seasons to check the potency of the plants.

8. The plants should not be affected bt insects or pests or ant microorganism which can lead to changes in the levels of secondary metabolites.

7.4. Identification of plant samples

The collected samples must be identified elsewhere.

1. Reviewing the flora of the region to compile a list of the plants that are in interest and to separate them from the plants that are to be avoided
2. Field identification must be done. They must at least be identified to their level of genus
3. To aid the identification, taxonomic experts should identify the plant species with a permanent scientific record or in case of a voucher specimen, the plant with the reproductive organs must be submitted to the major institutions or herbaria of the source country.
4. The voucher specimen is always collected as a twig with inflorescence in case of larger trees and shrubs. Whereas, herbs are collected as a whole plant with flowers and fruits.

7.5. Cleaning of plants

Proper cleaning of plants is an important step after collection. This process involves the following steps of washing, peeling, stripping leaves from the stems. This is usually done in hands to have better results.

7.6. Drying

The plant materials are dried to remove the water content and thus after the removal of water so that they can be stored. This process should be done immediately as soon as the plants are collected so that it is prevented from spoilage. There are two methods in drying the plants,

Natural process

This process include sun-drying. In this the plants are kept in the shades and are air dried in sheds. This process takes few weeks for complete drying of the moisture. This time depends on the temperature and humidity, in natural process there are two methods, they are sun- drying and shade drying.

Sun-drying

Sun-drying involves the drying of plant samples under direct sunlight; they are left in sun light for few days until they get dried and brittle. The temperature is usually about $33\pm4°C$ and may take few days for complete drying

Shade drying

This method involves drying the plant materials under sunlight exposure but not in direct sunlight in a good ventilated area. This is the most predominantly followed method for phytochemical analysis because they do not change the secondary metabolites of the plants. The temperature is around 31±4°C and will take a week for complete drying.

Artificial drying

Hot air oven drying

Artificial drying is done using the help of artificial driers. This process will reduce the time consumed to few hours or minutes. The common method used for the drying of medicinal plants is warm-air drying. This is done using the hot air oven on which warm air is blown. This method is applicable for drying of succulent parts of plants and fragile flowers. The drying must be done in lower temperature to prevent the thermolabile compounds being disintegrated.

Freeze drying

It is a process where an instrument called lyophilisor or freeze dryer is used to dry the plant samples. Both fresh and dried samples can be lyophilised and it is one of the modern methods and has very high advantages like, freezing the sample inactivates all enzymes and microbial activity, as well as hydrolytic compounds can also be stable. Any plant sample like leaf, stem, roots, bark, fruit or flowers can be lyophilised; the phytochemicals in the plant materials will not be degraded for several years when they are stored in freezer.

7.7. Grinding of plant materials

After complete drying of moisture the plant samples are to be powdered for the further analysis. There are different types of powdering, they include the following

1. Grinding can be done by milling in an electric grinder or by a spice mill or can also be in mortar or pestle.
2. Grinding increases the efficiency of the extraction due to increased surface area of the plants. The decrease in the surface area can lead to dense packing of the material.
3. Milling the plants into a fine powder is always ideal but if they are too fine this affects the solvent's flow and also produces

more heat which could degrade some thermolabile compounds.

4. Ball mining can also be done where the size of the powdered components are usually of same size.

7.8. References

• Darfour BernardAsare Isaac Kwabena1 , Ofosu Daniel Osei , G. Achel Daniel , S. Achoribo Elom and Agbenyegah Sandra The Effect of Different Drying Methods on the Phytochemicals and Radical Scavenging Activity of Ceylon Cinnamon (Cinnamomum zeylanicum) Plant Parts. European Journal of Medicinal Plants 4(11): 1324-1335, 2014

• Vishnu Balamurugan, Sheerin Fatima .M.A, and Sreenithi Velurajan. "A GUIDE TO PHYTOCHEMICAL ANALYSIS" International Journal Of Advance Research And Innovative Ideas In Education Volume 5 Issue 1 2019 Page 236-245

• Sivanandham, Velavan. (2015). PHYTOCHEMICAL TECHNIQUES - A REVIEW. World Journal of Science and Research. 1. 80-91

Chapter eight

Choice of solvents

8.1. Introduction

The solvent that is being used for the extraction process is very important in determining the biologically active phytochemicals from the plants. These solvents must be less toxic, easy to evaporate in less heat, should preserve the compounds in it and should not dissociate it. The various solvents commonly used for extraction include water, alcohol, acetone, chloroform and ether.

8.2. Water

Water is a universal solvent; but they are only used for the extraction of polar compounds so choosing water as a solvent depends on the nature of the compound to be extracted. Plant extracts with anti-microbial activities are usually extracted with water. But the organic solvents give consistent results in anti-microbial activities when compared to water. Water soluble compounds in the extract cannot give significant results. Water can be used for the isolation of pigments

8.3. Alcohol

These alcoholic extracts of plants show more activity than aqueous extracts due to the presence of higher amounts of polyphenols. This is because of the higher cell wall and seed degradation by the alcohols which releases the polyphenols which will be degraded in the case if aqueous extracts. But Ethanol is more microbicidal than water. More bioactive compounds are extracted in 70% ethanol than pure ethanol. Ethanol is also found easier to extract intracellular ingredients from plant materials. Polar solvents like methanol, ethanol and their aqueous mixtures are used for extraction of phenolic compounds. Addition of water to alcohol will improve the rate of extraction. Methanol is more polar but it is unsuitable for extraction due to its cytotoxic nature.

8.4. Acetone

Acetone dissolves many hydrophilic and lipophilic compounds from the plants and it is miscible with water. It is low toxic and volatile and it is used for extracting antimicrobial activities. Extracting tannins and other phenolic compounds are done with acetone. They are also used to extract saponins.

8.5. Chloroform

Terpenoid lactones are obtained from barks by extraction with chloroform. Tannins and Terpenoids are treated with less polar solvents.

8.6. Ether

They are used for the extraction of coumarins and fatty acids.

8.7. Reference

- Vishnu Balamurugan, Sheerin Fatima .M.A, and Sreenithi Velurajan. "A GUIDE TO PHYTOCHEMICAL ANALYSIS" International Journal Of Advance Research And Innovative Ideas In Education Volume 5 Issue 1 2019 Page 236-245

Chapter nine

Methods of extraction

9.1. Introduction

The processes of extraction involve isolation of primary, secondary metabolites and other important nutrients from the plant cells after lysis of them. There are several types of extraction processes involved in phytochemical analysis and some of the important extraction processes are Homogenization, serial exhaustive extraction, Soxhlet extraction, maceration, decoction, infusion, digestion, percolation and sonication

9.2. Homogenization

This method is one of the most widely used methods for extraction. This is either done by dried or wet extraction method. In this dried extraction method the dried plant samples are finely powdered and added to the solvent mixed for few minutes and kept in an orbital shaker for about 24 hours. In wet extraction process, the parts of the plants are cut into small pieces, grinded in a mortar and pestle and are added to a solvent and shaken in an orbital shaker for 24 hours and then filtered. The filtrate can be used for the further analysis.

9.3. Serial Exhaustive Extraction

It is done with a variety of solvents from a non-polar solvent like hexane to more polar solvent like methanol to extract a wide polarity range of compounds. The disadvantage is that thermolabile compounds cannot be extracted due to the high heat which leads to the degradation.

9.4. Soxhlet Extraction

It is used when the compound is less soluble in the solvent and the impurities are soluble in the solvent. If the desired compound is highly soluble in the solvent the impurities can be removed by simple filtration. The advantage is that the solvent is recycled in this method and hence there is less wastage of the solvent. Similar to the above method, thermolabile compounds cannot be extracted in this method

9.5. Maceration

In this method, whole plant or the powder can kept in the solvent for a certain period with frequent agitation until the soluble compounds are dissolved. This method is the best suitable method for the thermolabile compounds

9.6. Decoction

In this method heat stable and water soluble compounds are extracted. This is extracted plant materials are boiled in the water for about 15 minutes and are cooled, filtered and are used for further analysis

9.7. Infusion

It is done by diluting the compounds in the solvents. It is prepared by macerating the compounds for a short period in cold or boiling water

9.8. Digestion

This is a process where the extraction is done as maceration with a gentle heat applied. It is used when the elevated temperature do not interfere the solvent efficiency or the compounds.

9.9. Percolation

For this process an instrument called percolator is used which is a narrow, cone shaped Vessel with open ends. The ingredients are moistened with an appropriate amount of the specified menstrum and allowed to stand for approximately 4 h in a well closed container, after which the mass is packed and the top of the percolator is closed. Additional menstrum is added to form a shallow layer above the mass, and the mixture is allowed to macerate in the closed percolator for 24 h. The outlet of the percolator then is opened and the liquid contained therein is allowed to drip slowly. Additional menstrum is added as required, until the percolate measures about three quarters of the required volume of the finished product. The marc is then pressed and the expressed liquid is added to the percolate. Sufficient menstrum is added to produce the required volume, and the mixed liquid is clarified by filtration or by standing followed by decanting.

9.10. Sonication

In this method the ultrasound with higher frequencies of 20 kHz – 2000 kHz are used which will disrupt the cells and releases the constituents. Although the process is useful in some cases, like extraction of rauwolfi a root, its large-scale application is limited due to the higher costs. One disadvantage of the procedure is the occasional but known deleterious effect of ultrasound energy (more than 20 kHz) on the active constituents of medicinal plants through formation of free radicals and consequently undesirable changes in the drug molecules.

9.11. References

- Ashis, (2003) Herbal falk remides of Bankura and medinipur districts, west Bengal.Indian Journal of Traditional knowledge 2 (4) :393-396.

- Vishnu Balamurugan, Sheerin Fatima .M.A, and Sreenithi Velurajan. "A GUIDE TO PHYTOCHEMICAL ANALYSIS" International Journal Of Advance Research And Innovative Ideas In Education Volume 5 Issue 1 2019 Page 236-245

- Sivanandham, Velavan. (2015). PHYTOCHEMICAL TECHNIQUES - A REVIEW. World Journal of Science and Research. 1. 80-91

- Das K, Tiwari RKS, Shrivastava DK. Techniques for evaluation of medicinal plant products as antimicrobial agent: Current methods and future trends. Journal of Medicinal Plants Research 2010; 4(2): 104-111.

- Ncube NS, Afolayan AJ, Okoh AI. Assessment techniques of antimicrobial properties of natural compounds of plant origin: current methods and future trends. African Journal of Biotechnology 2008; 7 (12): 1797-1806.

Chapter ten

Qualitative and quantitative analysis of primary metabolites

10.1 Introduction

A primary metabolism defined as the biological reactions that are essential for the survival of a living organism. The intermediate end products of the primary metabolism are termed as primary metabolites i.e. Most of that carbon, nitrogen, and energy ends up in molecules that are common to all cells and are required for the proper functioning of cells and organisms. These molecules, e.g., carbohydrates, proteins, nucleic acids, and lipids are called primary metabolites. In this chapter we will be seeing about the qualitative and quantitative analysis of primary metabolites.

10.2 Qualitative analysis of primary metabolites

Test for carbohydrates

Test for sugars

1. Benedict's test: About 0.5 ml of the filtrate was taken to which 0.5 ml of Benedict's reagent is added. This mixture was heated for

about 2 minutes in a boiling water bath. The appearance of red precipitate indicates the presence of sugars

2. Molisch's test: To about 2ml of the sample, 2 drops of alcoholic solution of α-napthol was added and to the mixture after being shaken well. Few drops of conc.H2SO4 were added along the sides of the test tube. A violet ring indicates the presence of sugars

3. Fehling's Test: To about 1 ml of sample, 1 ml of Fehling's A solution (7 gm CuSO4.5H2O dissolved in distilled water containing 2 drops of dilute sulfuric acid) and 1 ml of Fehling's B (35 gm of potassium tartarate and 12 gm of NaOH in 100 ml of distilled water) solution are added and boiled for few minutes reddish brown precipitate is end point which indicates the presence of reducing sugars.

4. Barfoed's test: To 1 ml of sample, 2-3 ml of Barfoed's reagent (a solution of cupric acetate and acetic acid) is added and boiled for 5 minutes. Reddish precipitate indicates the presence of reducing sugars

5. The Bial's Test: To 1 ml of sample, 3 ml of Bial's reagent (orcinol, HCl and ferric chloride)

is added and heated over a Bunsen burner or a boiling water bath. Bluish green product indicates pentoses and brown or grey product indicates hexoses.

6. The Seliwanoff's Test: 1 ml of sample is taken and 3 ml of Seliwanoff's reagent (solution of resorcinol and HCl) is added and boiled, red colour is positive for ketoses.

7. The Hydrolysis Test: To 5 ml of sample, few drops of Conc. HCl is added and boiled orange colour indicates presence of sucrose.

8. The Osazone Test: to 2 ml of sample, 1.5 gram of phenylhydrazine hydrochloride, 0.6 gm of sodium acetate, and 4 mL of water are added and are boiled. After about half an hour, the precipitates are carefully taken and observed under microscope. The formation of needle shaped crystals is positive for Glucose/fructose, hedgehog shaped are positive for lactose and sunflower shaped are positive for maltose.

Test for Starch

1. To about 5 ml of distilled water, 0.01g of iodine and 0.075 g of potassium iodide were added and this solution was added to about

2-3 ml of the extract. Formation of blue colour indicates the presence of starch

2. Iodine test: few drops of iodine solution are added to the extract appearance of blue colour indicates the presence of starch

Test for proteins

1. Biuret test: 2ml of filtrate was taken to which 1 drop of 2% copper sulphate solution was added; 1ml of 95% ethanol was added. Then it was followed by excess addition of KOH. The appearance of pink colour indicates the presence of protein.

2. 2ml of extract was mixed with 2ml of water and about 0.5% of conc. HNO_3 was added. The appearance of yellow colour indicates the presence of proteins.

3. To about 2 ml of the extract, 2ml of miller's reagent was added white precipitate which turns red on heating will confirm the presence of proteins.

Test for amino acids

1. To 1ml of the extract, few drops of ninhydrin reagent (10mg of ninhydrin in 200ml of acetone) was added. The appearance of purple colour indicates the presence of amino acids.

2. To 2ml of extract few drops of nitic acid were added along the sides of the tube the appearance of yellow colour indicates the presence of protein and free amino acids.

Test for fatty acids

1.1 ml of the extract was mixed with 5 ml of ether. There extracts were allowed to evaporate on a filter paper and the filter paper was dried. The appearance of transparency indicates the presence of fatty oils

10.3. Quantitative analysis of primary metabolites

Estimation of carbohydrates

The qualitative analysis of sugars can be determined by the methods in qualitative analysis by establishing a standard curve and comparing them with the test samples.

Anthrone method: 1ml of the test sample and different concentration of standard are taken in a test tube to which 4 ml of anthrone reagent (200mg of anthrone reagent in 100ml of concentrated H_2SO_4) was added. Incubated in water bath for few minutes and absorbance was read at 630nm against reagent blank and plotted against standard.

Estimation of proteins

1. Lowry's method: different concentration of bovine serum albumin (standard) and test solutions are taken and the final volume is made upto five ml using distilled water. To about 0.5 ml of protein solutions, 2 ml of alkaline copper sulphate solution* is added and incubated in room temperature for 15 minutes. Then 0.2 ml of Folin-Ciocalteau solution is added and incubated for 30 minutes. The Optical density (absorbance) is measured at 660nm against reagent blank and plotted against standard graph.

* Alkaline copper sulphate: **Solution 1** -50 ml of 2% sodium carbonate mixed with 50 ml of 0.1 N NaOH solutions (0.4 gm in 100 ml distilled water , **solution 2-** 10 ml of 1.56% copper sulphate solution mixed with 10 ml of 2.37% sodium potassium tartrate solution. Prepare analytical

reagents by mixing 2 ml of solution 2 with 100 ml of solution 1.

2. Bradford's method: 100 mg Coomassie Brilliant Blue G-250 was dissolved in 50 mL 95% ethanol to which 100mL of phosphoric acid was added and made up to 1 litre, filtered and stored in a brown bottle at 4 ∘C. 0.1 ml of various concentration of standards and test solutions were taken and 5 ml of Bradford's reagent is added, incubated and absorbance was measured at 585nm against reagent blank.

3. The Kjeldahl Method: About 1 gram of the sample is taken and 15ml of conc. H_2SO_4 was added with two copper crystal tablets in a heat block at 420 ∘C and heated for 2 hours. After cooling, water was added and neutralised and titrated. The total amounts of nitrogen released are multiplied with traditional conversion factor of 6.25 and species specific conventional factor to determine protein content.

Estimation of fats

Two gram of the plant material is taken in a thimble which was prepared with whatman no. 1 filter paper and the mouth was closed with fat free absorbent cotton. 250ml Soxhlet flask was taken and the solvent was added to the flask till its neck. The sample was introduced into the soxhlet and the fat was extracted for 8 hours in 55 - 60 °C. After extraction, solvent is recovered. The receiver flask will contain approximately 25 ml of solvent and fat. The solvent was removed by evaporating it in a hot air oven at 95 °C. The fat was weighed and the amount was calculated.

Estimation of amino acids

Ninhydrin method: 1 ml of test sample and standards are taken in a test tube and 1 ml of ninhydrin reagent (8g of ninhydrin and dissolve in 100ml of acetone) was added to it. Incubated in boiling water bath for half an hour, 5 ml of diluent solution (equal volume of water and n-propanol) was added and incubated in room temperature for about 15 minutes before reading the absorbance at 570nm against reagent blank.

10.4. References

• Experimental Biochemistry, A student Companion by Beedu Sashidhar Rao and Vijay Deshpande.

• Sadasivam, S. and Theymoli Balasubramanian (1985). Practical Manual

(Undergraduate), TamilNadu Agricultural University, Coimbatore, p. 2.

•	Qualitative testing for carbohydrates by James O. Schreck and William M. Loffredo, Modular Laboratory Programme in Chemistry.

•	Lowry, O.H., Rosebrough, N.J., Farr, A.L., and Randall, R.J. (1951) J.Biol.Chem 193: 265 (The original method).

•	Hartree E.E. (1972). Anal. Biochem. 48:422 (This modification makes the assay linear over a larger range than the original assay)

•	Wilson, K. and Walker, J. (2000) "Practical Biochemistry: Principles and Techniques", Cambridge University Press.

•	Bradford, M. M. (1976). A rapid and sensitive method for the quantitation of microgram quantities of protein utilizing the principle of protein-dye binding. Anal Biochem 72: 248-254.

•	Stoscheck, C. M. (1990). Quantitation of protein. Methods Enzymol 182: 50-68.

•	Hanne K. Mæhre , Lars Dalheim, Guro K. Edvinsen, Edel O. Elvevoll and Ida-Johanne

Jensen. Protein Determination—Method Matters. Foods 2018, 7, 5; doi:10.3390/foods7010005

• Latimer, G.W. Official Methods of Analysis of AOAC International; AOAC International: Gaithersburg, MD, USA, 2016

• Kjeldahl, J. Neue Methode zur Bestimmung des Stickstoffs in organischen Körpern. Fresenius' J. Anal. Chem. 1883, 22, 366–382.

• Lourenço, S.O.; Barbarino, E.; De-Paula, J.C.; Pereira, L.O.d.S.; Lanfer Marquez, U.M. Amino acid composition, protein content and calculation of nitrogen-to-protein conversion factors for 19 tropical seaweeds. Phycol. Res. 2002, 50, 233–241.

• Moore S, Stein WH. Photometric methods for use in the chromatography of amino acids. J Bio Chem 1948;176:367-88.

• Hedge JE, Hofreiter BT. In: Whistler RL, Be Miller JN, editors.Carbohydrate Chemistry. Vol. 17. New York: Academic Press; 1962. p. 1-19.

• Mariotti, F.; Tome, D.; Mirand, P.P. Converting nitrogen into protein—Beyond 6.25

and Jones' factors. Crit. Rev. Food Sci. 2008, 48, 177–184.

Chapter eleven

Qualitative analysis of miscellaneous compounds

11.1 Introduction

There are some other substances other than primary and secondary metabolites, these miscellaneous compounds are found alongside with the secondary metabolites and are similar to them. The important miscellaneous compounds are Carboxylic acids, Resins, fixed oils and fats, Gums and mucilage. The qualitative analyses of these miscellaneous compounds are following.

11.2. Test of resins

1.   Precipitation test: about 0.2 g of extract was extracted with 15ml of 95% ethanol. The alcoholic extract was then poured into a beaker containing about 20ml of distilled water.

2.   1ml of extract was taken and to this few ml of acetic anhydride was added to this 1ml of conc.H^-2SO_4 was added. The appearance of orange to yellow colour indicates the presence of resins

106

11.3. Test of fixed oils and fats

1.   Spot test: small quantity of the extract was taken and pressed between 2 filter papers. The appearance of spots indicates presence of oils

2.   Saponification test: To the extract, few drops of 0.5N alcoholic KOH and few drops of phenolphthalein were added. This mixture was heated for about 2 hours. The formation of soap or partial neutralization of alkali indicates the presence of fixed oils or fats

11.4. Gums and mucilage

To 1ml of extract, distilled water, 2ml of absolute ethanol was added with constant stirring white or cloudy precipitate indicates the presence of gums or mucilage

11.5. Carboxylic acids

1.   To 1ml of extract a pinch of sodium bicarbonate is added. The production of effervescence indicates the presence of carboxylic acids

2.   2ml of alcoholic extract was taken in warm water and filtered. The filtrate was then tested with litmus paper and methyl orange. The

appearance of blue colour indicates presence of acids.

11.6. References

- Vishnu Balamurugan, Sheerin Fatima .M.A, and Sreenithi Velurajan. "A GUIDE TO PHYTOCHEMICAL ANALYSIS" International Journal Of Advance Research And Innovative Ideas In Education Volume 5 Issue 1 2019 Page 236-245

Chapter twelve

Qualitative and quantitative analysis of secondary metabolites

12.1 Introduction

Secondary metabolites are the compounds that are not utilized in the growth and development of the plant but are required for the survival in its environment. The secondary metabolites are used in the communication of the organisms to its mutualistic and also helpful in protecting them against pathogens and other antagonistic interactions. These are also helpful in avoiding stress tolerance, also as anti-microbial agents etc. The qualitative and quantitative analyses of some important secondary metabolites are as following.

12.2. Qualitative analysis secondary metabolites

Test for anthraquinones

To 5ml of extract, few ml of conc.H_2SO_4 was added and 1ml of diluted ammonia was added to it.

The appearance of rose pink confirms the presence of anthraquinones

Test for quinones

To 1ml of extract, alcoholic KOH is added the presence of red to blue colour indicates the presence of quinones

Test for alkaloids

      1.   Mayer's test: to a few ml of filtrate, 2 drops Mayer's reagent was added a creamy or white precipitate shows a positive result for alkaloids.

      2.   Wagner's test (iodine – potassium iodine reagent): To about an ml of extract few drops of Wagner's reagent were added. Reddish – brown precipitate indicates presence of alkaloids.

      3.   To 5ml of extract 2ml of HCl was added. Then 1 ml of Dragendroff's reagent was added an orange or red precipitate shows a positive result for alkaloids.

Test for glycosides

      1.   Borntrager's test: to 2ml of filtrate, 3ml of chloroform is added and shaken. The

chloroform layer is separated and 10% ammonia solution was added. The pink colour indicates the presence of glycosides

2.   5ml of extract was hydrolysed with 5ml of conc. HCl boiled for few hours in a boiling water bath, small amount of alcoholic extract was dissolved in 2ml of water and 10% of aqueous 10% NaOH was added the presence of yellow colour was a positive result for the glycosides.

3.   2ml of extract is mixed with about 0.4 ml of glacial acetic acid containing traces of ferric chloride and 0.5 of conc. H_2SO_4 was added the production of blue colour is positive for glycosides.

Test for cardiac glycosides (Keller-Killani test)

1.   5ml of solvent extract was mixed with 2ml of glacial acetic acid and a drop of ferric chloride solution was added followed by the addition of 1ml of conc. H_2SO_4. A brown ring in the interface indicates the presence of deoxy sugars of cardenoloides. A violet ring may appear beneath the brown ring while acetic acid layer a green ring may also form just gradually towards the layer.

Test for phenol

1.   Gelatine test: To 5ml of extract 2ml of 1% solution of gelatine containing 10% of NaCl is added. Appearance of white precipitate indicates the presence of phenol

2.   Lead acetate test: To 5 ml of extract 3ml of 10%lead acetate solution was added and mixed gently. The production of bulky white precipitate is positive for phenols.

Test for polyphenols

1.   To the 3ml of extracts 10ml of ethanol was added and were warmed in a water bath for 15 minutes. To this few drops of ferric cyanide (freshly prepared) was added. The formation of blue – green colour indicates presence of polyphenols.

2.   To 1ml of extract few drops of 5% solution of lead acetate was added. The appearance of yellow precipitate indicates the positive results for polyphenols

3.   To the 5ml of ethanolic extract 3ml of 0.1% gelatine solution was added. The formation of precipitate was positive for polyphenols

Test for tannins

1.   To 5ml of extract few drops of neutral 5% ferric chloride solution was added, the production of dark green colour indicates the presence of tannins

Test for Flavonoids

1.   To the aqueous solution of the extracts 10% ammonia solution is added and is heated. The production of fluorescence yellow is positive for flavonoids.

2.   1ml of extract was taken and 10% of lead acetate was added. The yellow precipitate is positive inference for the flavonoids

3.   The extract is treated with concentrated H_2SO_4 resulting in the formation of orange colour indicates the positive result for flavonoids.

4.   To 5ml of dilute ammonia the plant extract is added and shaken well. The aqueous portion is separated and concentrated H_2SO_4 is added. The yellow colour indicates the presence of flavonoids.

Test for phytosterols

1.   The extract is dissolved in 2ml of acetic anhydrite and to which 1 or 2 drops of concentrated H_2SO_4 is added along the sides an array of colour change indicates the presence of phytosterols.

2.   The extract was refluxed with alcoholic KOH and saponification takes place. The solution was diluted with ether and the layer was evaporated and the residue was tested for phytosterols. It was dissolved in diluted acetic acid and few drops of concentrated H_2SO_4 are added. The presence of bluish green colour indicates the presence of phytosterols.

Test for phlobatannins

1.   Aqueous extract was boiled with diluted HCl leading to the deposition of reddish precipitate indicates the presence of phlobatannins

Test for saponins

1.   0.5 mg of extract was vigorously shaken with few ml of distilled water. The formation of frothing is positive for saponins

2. The froth from the above reaction is taken and few drops of olive oil is added and shaken vigorously and observed for the formation of emulsion.

Test for steroids

2ml of extract with 2ml of chloroform and 2ml of concentrated H_2SO_4 are added, the appearance of red colour and yellowish green fluorescence indicates the presence of steroids

Test for xanthoproteins

1ml of extract is taken and to this few drops of nitric acid and ammonia are added. Reddish brown precipitate indicates the presence of xanthoproteins

Test for chalcones

2ml of ammonium hydroxide is added to 0.5 g of extract. The appearance of red colour indicates the presence of chalcones

Test for Terpenoids (Salkowski test)

3ml of the extract was taken and 1ml of chloroform and 1.5 ml of concentrated H_2SO_4 are added along the sides of the tube. The reddish brown colour in the interface is considered positive for the presence of terpenoids

Test for triterpenoids

To 10 mg of extract 1ml of chloroform is added and is mixed to dissolve it. 2ml of concentrated H_2SO_4 is added followed by 1ml of acetic anhydride. Formation of reddish violet colour is positive for the presence of triterpenoids.

Test for anthocyanins

2ml of aqueous extract was taken to which 2N HCl was added and it was followed by the addition of ammonia, the conversion of pink-red turns blue-violet indicates the presence of anthocyanins.

Test for Leucoanthocyanins

To 5ml of extract dissolved in water, 5ml of Isoamyl alcohol is added. The red appearance of the upper layer indicates the presence of Leucoanthocyanins

Test for Coumarins

To 2 ml of the extract, 3 ml of 10% aqueous solution of NaOH is added. The production of yellow colour indicates the presence of coumarins

Test for emodins

To 5ml of extract, 2ml of NH_3OH and 3ml of benzene are added. The production of red colour indicates the presence of emodins

12.3. Quantitative analysis of secondary metabolites

1. Total phenolic content

Gallic acid is used as standard. 1 ml of different concentrations of standards and test solutions are taken and to them 1 ml of Folin-Ciocalteau reagent was added incubated for 5 minutes and 1 ml of 10% sodium carbonate was added and incubated for half an hour. The absorbance was read at 730nm against reagent blank.

2. Total alkaloid content

1. Harborne method: To about 10 grams of homogenised plant material, 20 ml of methanol: ammonia (68:2) was added and the solution was replaced with fresh solution of Methanolic ammonia. The procedure was repeated three times and the extract was evaporated by using a flash evaporator. The residue was treated with 1N hydrochloric acid for overnight and was extracted with ammonia trice and the organic layer was

pooled and evaporated to dry them. The acidic layer was basified with NaOH to pH of 12 and extracted with chloroform trice, and then the chloroform layer was pooled and dried by evaporation. The fraction was weighed and alkaloid content was expressed as mg/100 g.

2. 50 mg of extract was dissolved in 100ml of 2N HCl and was filtered; it was then transferred to a separation funnel. 5 ml of Bromocresol green solution (69.8 mg Bromocresol green + 3 ml of 2N NaOH dissolved in 5 ml distilled water and made upto 1000ml) and 5 ml of phosphate buffer (pH 4.7) were added. The mixture was shaken vigorous and extracted as 2ml aliquots and diluted with chloroform. Similarly atropine standards are treated and different concentrations of these standards are also taken. The absorbance was read at 470nm in a UV- Vis Spectrometer against reagent blank.

3. Total flavonoid content

1. 10 mg of Quercetin is used as standard; it was diluted in 50 ml of methanol to a concentration of (200µg/ml) similarly, 2.5 ml of sample extract was dissolved in methanol at 1mg/ml concentration. Different volumes of both

test and standard solutions are taken and 0.1 ml of 1M potassium acetate solution is added. Incubated for 6 minutes and 0.1 ml of 10% aluminium chloride solution is added and incubated for 30 minutes and read at 415 nm against reagent blank.

2. 1ml of the extract was mixed with 0.075 ml of 5% sodium nitrate solution and incubated at room temperature for 5 minutes and to it 10% $AlCl_3$ solution was added and incubated for 6 minutes then few drops of 1N NaOH was added and was measured at 510nm against reagent blank.

4. Determination of tannins

1 ml of sample was mixed 5 ml of vanillin hydrochloride reagent (Mix equal volumes of 8% hydrochloric acid in methanol and 4% vanillin in methanol). Incubated for 20 minutes in room temperature and was measured at 500nm against reagent blank.

5. Determination of steroids

Dissolved 50 mg of the standard, Lupeol 50ml of chloroform (1000 µg/ml) similarly; 25mg of sample was dissolved in n-Hexane and filtered. Different concentrations of the standard and test solutions are taken and 1 ml of chloroform is added and the chloroform was evaporated, 0.5 ml of glacial acetic acid is added and 5 ml of Liebermann Burchard's reagent (5 mL of acetic anhydride and 5 mL of concentrated sulfuric acid are added into 50 mL absolute ethanol, while cooling in ice and then heated at 100° C for 5-10 minutes) was added and made into 10ml using chloroform. It was then Incubated for 20 minutes, read at 618nm against reagent blank.

12.4. References

• Barkat M. Z., Shehab S. K., Darwish N., Zahermy E. I., (1973) Determination of ascorbic acid from plants. Analyst Biochem; 53: 225-245.

• Bimakr M. Comparison of different extraction methods for the extraction of major bioactive flavonoid compounds from spearmint (Mentha spicata L.) leaves. Food Bioprod Process 2010; 1-6.

• Nikhal SB, Dambe PA, Ghongade DB, Goupale DC. Hydroalcoholic extraction of Mangifera indica (leaves) by Soxhletion. International Journal of Pharmaceutical Sciences 2010; 2 (1): 30-32.

• Remington JP. Remington: The science and practice of pharmacy, 21st edition, Lippincott Williams & Wilkins, 773-774.

• Handa SS, Khanuja SPS, Longo G, Rakesh DD. Extraction Technologies for Medicinal and Aromatic Plants. International centre for science and high technology, Trieste, 2008, 21-25.

• Evans.W.C, "Treaseand Evans Pharmacognosy", Harcourt Brace and company. Asia pvt. Ltd.Singapore, 1997.

• Hasler CM and Blumberg JB (1999) Phytochemicals: Biochemistry and physiology. Introduction. Journal of Nutrition 129: 756S–757S.

• Hazra, K. M., Roy R. N., Sen S. K. and Laska, S. (2007). Isolation of antibacterial pentahydroxy flavones from the seeds of Mimusops elengi Linn. Afr. J. Biotechnol. 6 (12): 1446-1449.

• Harborne JB. Phytochemical Methods. London: Chapman and Hall, Ltd.; 1973. p. 49-188.

• Khandelwal, K.R (2006) Practical Pharmacognosy (16th ed.,) Nirali Prakashan, Pune.p98-106.

• Kokate C.K, Practical Pharmacognosy. Pune : Vallabh Prakashan;2003.

• Kumar S. et al., Antioxidant free radical scavenging potential of Citrullus colocynthis (L.) Schrad. Methanolic fruit extract, Acta pharma. 2008, 58:215- 220.

• Lapornik B, Prosek M, Wondra, A. G. Comparison of extracts prepared from plant by-products using different solvents and extraction time. Journal of Food Engineering 2005; 71: 214–222.

• Mabry TJ, Markham KR, Thomas MB. The systematic identification of flavonoids. New York: Springer Publishers;1970. p. 84-88.

• Markham KR. Techniques of flavonoid identification. London:Academic Press; 1982.

• Mathai K(2000). Nutrition in the Adult Years. In Krause"s Food, Nutrition, and Diet Therapy, 10th ed., ed. L.K. Mahan and S. Escott-Stump; 271: 274-275.

• Obadoni BO, Ochuko PO (2001). Phytochemical studies and comparative efficacy of the crude extracts of some Homostatic plants in

Edo and Delta States of Nigeria. Global J. Pure Appl. Sci. 8 b:203-208.

• Okwu DE. Phytochemicals and vitamin content of indigenous spices of Southeastern Nigeria. J. Sustain. Agric. Environ. 2004; 6 (1): 30- 37.

• Geedhu daniel, krishnakumari s. Quantitative analysis of primary and secondary metabolites in aqueous hot extract of Eugenia uniflora (L.) Leaves. Asian J Pharm Clin Res, Vol 8, Issue 1, 2015, 334-338.

• D Naik, Pralhad & Mishra, Rajnarayan. (2013). Quantitative Analysis of Secondary Metabolites of Withania Somnifera and Datura Stramonium. International Journal of Science and Research (IJSR). 4. 4-438

• Patel KK. (2005) Master dissertation. Shorea robustra for burn wound healing and antioxidant activity. Department of Pharmacology, KLESS College of Pharmacy, Karnataka, India, p.33.

• Jia Z, Tang M, Wu J. The determination of flavonoid content in mulberry and their scavenging effects on superoxide radicals. Food Chem 1999;64(4):555-99.

• Bray HG, Thorpe WV. Analysis of phenolic compounds of interest in metabolism. Methods Biochem Anal 1954;1:27-52.

Chapter thirteen

Qualitative and quantitative analysis of vitamins

13.1. Introduction

The term vitamin means vital amine and was coined by Casimir Funk in 1912. The vitamins are important for the human health and they have central roles in the metabolism. As vitamins are a part of our balanced dirt, it is necessary to take a lot of fresh fruits and vegetables as they have been rich sources of vitamins. As they have important metabolic functions in animals they must also have some important function in their origin too and they play an important role in plant metabolism too. Qualitative and quantitative analysis of vitamins are as following.

13.2. Qualitative analysis of vitamins

Test for Vitamin – A

In 5 ml of chloroform, 250mg of the powdered sample is dissolved and it is filtered, to the filtrate, 5ml of antimony trichloride solution is added. The appearance of transient blue colour indicates presence of vitamin-A

Test for vitamin – C

In 5ml of distilled water, 1ml of the sample was diluted and a drop of 5% sodium nitroprusside and 2ml of NaOH is added. Few drops of HCl are added dropwise, the yellow colour turns blue. This indicates the presence of vitamin- C

Test for vitamin – D

In 10 ml of chloroform, 500mg of powdered extract is dissolved and filtered. 10ml of antimony trichloride is added, the appearance of pinkish-red colour indicates the presence of vitamin – D

Test for vitamin – E

Ethanoic extract of the sample was made and filtered (500mg in 10ml), few drops of 0.1% ferric chloride were added and 1ml of 0.25% of 2'-2'dipyridyl was added to 1ml of the filtrate. Bright-red colour was formed with a white background.

13.3. Quantitative analysis of vitamins

Estimation of vitamin A

Neeld and Pearson method: 0.5 ml of the sample extract is mixed with 0.5 ml of chloroform and to which 2 ml of trifloro acetate reagent (Commercially available) is added. The absorbance was measured at 600nm against reagent blank.

Estimation of β- carotene

Neeld and Pearson method: 1 ml of sample and 1 ml of 90% alcoholic 2N KOH are added and was heated for 30 minutes. After cooling, 25 ml of distilled water was added and was transferred to a separating funnel. It was extracted trice with 25, 15, 10 ml of petroleum ether at a temperature of around 50 °C. It was then pooled and washed with distilled water until the alkali is removed. Then it was dried by adding sodium sulphate, 3 ml of petroleum ether phase was taken in a cuvette and measured in 420nm against reagent blank.

Estimation of vitamin B₁

Sadasivam and Manickam method: 5 gram of sample was added to 100ml of 0.1N H_2SO_4 taken in a conical flask and was slowly rotated overnight and was vigorously rotated the next day and was filtered. In a 100 ml separating funnel, 10 ml of the extract was taken and 3 ml of 15% sodium hydroxide and few drops of ferri cyanade were added and shaken for a minute. 15 ml of iso-butanol was added and was again shaken vigorously for a minute. The bottom layer was discarded and 0.2 sodium sulphate was added mixed gently and collected in test tubes. Similarly a blank was created without ferri cyanade. They are read on 366nm against blank and the thiamine content is calculated as

Thiamine (μg/100 g) =

[(0.25 x 10)/ a-a1] x [(b-b1) x 100/ 10] x [10/5]

Where, a - OD of standard;

a1 - OD of standard blank;

b – OD of sample;

b1 - OD of sample blank

Estimation of vitamin C

Omaye method: About 0.5 ml of sample extract was taken and 0.5 ml of distilled water was added to which 0.2 ml of DTCS reagent was also added.

It was incubated at room temperature for the development of crystals for 3 hours. 1.5 ml of ice cold H_2SO_4 is added, mixed and incubated for another 30 minutes. The OD at 520 nm was measured against reagent blank.

Estimation of vitamin E

Varley method: 1.5 ml of standard, test and blank were taken in centrifuge tubes. 1.5 ml of ethanol is added to test and blank solutions and 1.5 ml of water is added to standard. 1.5 ml of xylene is added and is centrifuge. 1.0 ml of xylene layer is taken and 1.0 ml of 2, 2'- dipyridyl reagent was added. The absorbance of test and standards are measured against blank at 460nm. Then to the tubes, 0.33 ml of ferric chloride is added and mixed. After exactly 15 minutes, the test and standards are again read at 520nm against. The amount of Vitamin E is calculated by the following formula,

Vitamin E (µg/g) =

$$\frac{(\Delta A520nm - \Delta A460nm) \times conc\ [S] \times 0.29 \times Total\ volume}{\Delta A520nm \times Vol\ for\ experiment \times Weight\ of\ sample}$$

13.4. Reference

• AOAC. (1984) Vitamins and other nutrients. In Official Methods of Analysis of the Association of Official Analytical Chemists. 14th Edition (William S, ed.), AOAC,Virginia; pp. 838 – 841.

• M Angeline Christie Hannah, S Krishnakumari. Profiling of Lipid and Vitamin contents in the extract of Watermelon (Citrullus vulgaris Schrad.). Journal of Pharmacognosy and Phytochemistry 2015; 4(3): 247-252

• Neeld JB Jr, Pearson WN. Macro- and micromethods for the determination of serum vitamin A using trifluoroacetic acid. J Nutr. 1963 Apr;79:454-62. doi:10.1093/jn/79.4.454. PubMed PMID: 13937886.

• Sadhasivam S, Manickam A. Biochemical methods for agricultural Science, 1996.

• Omaye ST, Turnball TD, Salberlich HE. Selected methods for the determination of ascorbic acid in animal cells, tissues and fluids method, Enzmol 1971; 62:1-11.

• Varley H. Practical Clinical Biochemistry; ArnoldHeinmann Publishers Pvt. Ltd 1976; 4:452.

Chapter fourteen

Qualitative and quantitative analysis of minerals

14.1. Introduction

Minerals have a very big role in the metabolism of plants; both the high concentration and low concentration of minerals in the plants have severe effects in their metabolism. These minerals are important for the biosynthesis of certain secondary metabolites which provide them antioxidant, antibacterial, antifungal, pesticidal and insecticidal activities. The contents of minerals in the soil vary and due to which there are several adverse effects in the metabolic activities of plants. Different stages of plants need different micro and macro nutrients for their growth and development. The qualitative and quantitative analyses of minerals are following.

14.2. Total ash content

Ash is the residue of the plant samples after incineration at 550-600 °C. For producing ash, the plant samples are incinerated at high temperatures of 600 °C. During this process due to high temperature, the organic compounds decompose and vaporise leaving inorganic elements in the ash.

Protocol

The procedure involves taking about two grams of sample in a vitrosil silica crucible. For about 30 minutes, the crucible was heated on a hot plate till the sample is black and charred. Then the crucible was heated in a muffle furnace at temperature upto 600°C for about 2 hours. The ash was cooled and weighed.

Water soluble ash (WSA)

The ash was boiled in distilled water (25ml) for five minutes, and then it was filtered and washed with hot water, ignited and weighed. The difference between the weight of ash and insoluble matter represents the water soluble ash.

Acid insoluble ash (AIA)

To the ash created above, 50 ml of 5N HCl is added and heated in a water bath for 30 minutes, and then it was cooled and filtered. Then filter paper is washed and dried along with the insoluble portion of the ash at 100 °C. The weight was determined and the acid soluble portion is used for mineral analysis.

14.3. Qualitative analysis Minerals

Ash material is prepared and 3:1 of nitric acid and hydrochloric acid are added to the material and allowed for about an hour. Then the solution is filtered and is used for mineral analysis.

Calcium

To about 10 ml of the above filtrate solution, one drop of dilute ammonium hydroxide solution and saturated ammonium oxalate solution are added white precipitates of calcium oxalate indicates the presence of calcium in the sample. These crystals are insoluble in acetic acid and are soluble in hydrochloric acid.

Magnesium

The calcium oxalate is removed from the above solution and was heated and then cooled; formation of white precipitate on addition of sodium phosphate indicates the presence of magnesium.

Potassium

To 2 ml of test solution, 2 drops of sodium cobalt nitrite solution is added. Presence of potassium is inferred by the formation of yellow precipitate of potassium cobalt nitrite.

Iron

To 5 ml of test solution, 2 drops of 2% ferrocyanide solution are added. Iron is confirmed by the production of dark blue colour.

Sulphate

Lead acetate is leaded to about 5 ml of test solution; white precipitate that can be dissolved in sodium hydroxide indicates the presence of sulphate

Phosphate

To about 5 ml of test solution, few drops of nitic acid and ammonium molybdate solution are added. It was then heated and cooled, yellow crystalline precipitate confirms the presence pf phosphates by precipitating yellow crystals of ammonium molybdate

Sodium

To 2 ml of test solution, few drops of uranyl magnesium acetate reagent are added. The yellow crystals of sodium magnesium uranyl acetate indicate the presence of sodium.

Chloride

To about 5 ml of test sample, 3 ml of lead acetate solution is added; white precipitate soluble in hot water confirms the presence of chlorine.

Nitrates

To about 5 ml of test solution, few drops of ferrous sulphate solution is added. Formation of brown ring in the clear solution while adding drops of sulphuric acid along the sides of the tubes indicates the presence of nitrates.

14.4. Quantitative analysis of Minerals

Nitrogen

This process for quantitative analysis of nitrogen involves two steps; they are digestion and distillation,

Digestion: In a Kjeldahls flask, 300mg of dry sample is taken and a pinch of catalyst mixture (1: 9: 0.02 of copper sulphate ($CuSO_4$), potassium sulphate (K_2SO_4) and Selenium dioxide (SeO_2)) is added. 7.5 ml of conc. Sulphuric acid is added slowly and heated until the acid boils and the mixture is clear (Apple green or colourless). The solution is made upto 50 ml using distilled water in a volumetric flask.

Distillation: Markham's steam distillation apparatus is used for the process of distillation. Stream is produced and a 50 ml conical flask containing 10 ml of boric acid solution (2% boric acid solution with indicator*), at the delivery end of condenser. The tip of the condenser is placed just beneath the surface of boric acid solution, 5 ml of previously digested solution is introduced into the distillation flask through a funnel and the funnel was closed with ground glass rod. 10 ml of 40% NaOH solution is added to the funnel and slowly introduced into the flask. Ammonia is formed along with the steam which was absorbed by the boric acid solution in the condenser outlet to form ammonium tetraborate ($(NH_4)_3BO_4$). Due to which the pink colour of the solution changes to green. The titration was continued till volume of 20ml. after distillation process, the ammonium tetraborate ($(NH_4)_3BO_4$) solution was titrated with 0.035N HCl** till the reappearance of pink colour which is the end point.

The strength of ammonia is calculated using the equation,

1 ml 0.035 N HCl = 0.5 mg of N

The result amount the amount of nitrogen equivalent to 300 mg of sample.

*indicator- 300 mg of bromocresol green and 200 mg methyl red was dissolved in 95% ethanol and made upto 500 ml with ethanol.

**0.035N HCl – 25 ml of conc. HCl is diluted with water and made upto 500ml (Stock solution). The normality of the solution is determined by titration with 1 g of ammonium tetra borate ((NH_4)3 BO_3) is taken in 50 ml conical flask with few drops of indicator. Normality is calculated with following formula.

$$\text{Normality of HCl} = \frac{1000}{\text{Titration value (ml)} \times 190.72}$$

After determining the normality of stock solution, 0.035 N HCl is prepared by appropriate dilution.

Calcium (Ca)

25 ml of Acid soluble ash portion was diluted to 150 ml using distilled water. To the solution, few

drops of methyl red indicator were added and the solution was neutralised with ammonia till the colour of solution changes from pink to yellow. Then the solution is heated and 10 ml of 6% solution of ammonium oxalate (COO.NH4). H2O) was added and boiled. Then it was again acidified by adding glacial acetic acid till the pink colour is appeared again. It was incubated for overnight in room temperature. The calcium oxalate crystals were filtered using Whatman filter paper No. 42 and was washed several times with water to remove acidity. It was then transferred to a beaker and 15 ml of 2 N H_2SO_4 was added and heated at 40°C and was titrated against 0.01 N $KMnO_4$ solutions until the pink colour as endpoint.

The amount of calcium is calculated using the equation.

$$1 \text{ ml of KMnO4} = 0.2004 \text{ mg of Ca}$$

Phosphorus (P)

0.5 ml of acid soluble solution was taken in a test tube (Volume can be changed according to the phosphorus content) and was diluted to 10 ml using distilled water. A blank is also made with water. To the test tubes, 1 ml of molybdate solution*, 0.4 ml of ANSA reagent ** were added and mixed. Incubated for five minutes and absorbance was measured at 660nm against blank. The amount can be determined by comparing with the standard phosphorus solution***.

* Molybdate solution – to 20 ml of distilled water, 25 g of ammonium molybdate was added and dissolved. 500 ml of 10N H_2SO_4 (200 ml concentrated H_2SO_4 to 520 ml of distilled water) was added and made upto 1 litre.

** Aminonaphtholsulfonic acid (ANSA) reagent – 195 ml of 15 % sodium bisulphite ($NaHSO_3$) was taken and to which 500 mg of 1, 2, 4 - aminonaphtholsulfonic acid was added and mixed. To this solution 5 ml of 20% sodium sulphite (Na_2SO_3) was added, mixed and stored in a brown bottle in 4°C

*** Standard phosphorus (P) solution – exactly 351mg of monopotassium phosphate (KH_2PO_4) was taken in a 1000ml standard flask and 500 ml of distilled water was added; to the solution, 10 ml of 10 N H_2SO_4 was added and volume was made upto 1000ml. five ml of the above solution contains 0.4mg of phosphorous

Potassium (K)

1 ml of the acid soluble portion was taken, diluted to 25 ml using distilled water. Distilled water was then fed to flame photometer atomizer in F.S control. The standard mixed solution was aspirated 1.7/0.8 mEq per litre on Na/K solution for 30 seconds. The F.S control was adjusted for Sodium as read out of 170 and potassium as 80 and was calibrated. The pressure of 0 to 10 mEq/1 and power 230 V + 10% 50 Hz is maintained. Then the prepared sample is fed to measure the relative concentration. Similarly, readings of stock standard of different concentration of 0.01 to 0.08 ml standard potassium* were made. The amount of potassium in sample is calculated by comparing it with standards.

*Standard potassium 10 mEq/litre (1 mEq/litre = 39 ppm). 0.746g of pure dry KC1 was dissolved in a litre of glass distilled water

14.5. Reference

• Sadasivam S, Manickam A. Biochemical methods for agricultural science, Wiley Eastern Ltd., Madras, 1996, 187–188. 28.

• Sivanandham, Velavan. (2015). PHYTOCHEMICAL TECHNIQUES - A REVIEW. World Journal of Science and Research. 1. 80-91.

• Raghuramalu N, Nair MK, Kalyanasundaram SA. Manual of laboratory techniques, second edition, national Institute of Nutrition, KMR, Hyderabad, 2003, 201. 29.

• Neil M, Neely M. Practical clinical chemistry, Ed Varley. H. Arnold – Heinzmann publishers Pvt. Ltd., 1956, 4, 465. 30.

• Fiske CJ, Subarrow Y. The colorimetric determination of phosphorus. J Biol Chem 1935, 66:375–400.

Chapter fifteen

Chromatographic methods of phytochemical analysis

15.1. Introduction

Chromatography is a technique used for separation of a mixture of compounds. Chromatography is based on the principle where molecules in mixture applied onto the surface or into the solid, and fluid stationary phase (stable phase) is separating from each other while moving with the aid of a mobile phase. A stationary phase is a solid or a liquid through which the mixture passes through and gets separated. The mobile phase can be a liquid or gas in which the mixture is dissolved. When the mobile phase pushes itself into the stationary phase the mixture gets separated into individual compounds. Then these separated compounds can be eluted from the stationary phase. Chromatography can be preparative or analytical, in phytochemical analysis; analytical techniques of chromatography are used.

Chromatographic methods have high resolving power and are capable of precise separation and identification of compound. Even extremely small quantities of compounds can be separated using these chromatographic techniques. Detectors like mass spectroscopy facilitates identification of the separated fractions which are detected and recorded both qualitatively and quantitatively making them a very significant and powerful method in qualitative analysis of plant compounds. High Performance Liquid Chromatography is applicable for compounds soluble in solvents. High performance thin layer chromatography is applicable for the separation, detection, qualitative and quantitative analysis of phytochemicals.

There are many different methods of chromatography used for the analysis of plant compounds some of the important methods include, Column chromatography, Paper Chromatography, Thin layer chromatography, Gas Chromatography, High Performance Liquid Chromatography (HPLC), High Performance Thin Layer Chromatography (HPTLC), Optimum Performance Laminar Chromatography (OPLC) and they are briefly discussed in this chapter.

15.2. Column chromatography

This method was first described by Matteucci which was later explained by Kuhn and Lederer by separating polyene pigments. The column chromatography involves the adsorption of solutes of the solution through a stationary phase which will separate this mixture of components into individual compounds. The separation is based on the affinity of the molecules towards the solvent (mobile phase) and the stationary phase. The molecules that have low affinity towards stationary phase elute first and those which have high affinity towards stationary phase elute later. They can be collected as individual components as they elute. This is one of the earlier methods and is still used till this date but there are several other modern methods that are faster and less expensive than column chromatography.

15.3. Paper Chromatography

One of the earliest discovered chromatography method by Richard Martin and was used by Mikhail Semenovich Tsvett. In this method, cellulose in the filter paper act as a stationary phase and a mixture of solvents can be used as mobile phase. This method is mainly used for analysis, identification, purification and quantification of individual components from a mixture. There samples trend to move along with the solvent (mobile phase) in the paper which acts as a stationary phase and gets separated in the paper, the substances are identified using the retention factor (R_f values) which was calculated using the formula,

$$R_f = \frac{\text{Distance travelled by the solute}}{\text{Distance travelled by the solvent}}$$

They are not mostly used for analysis of phytochemicals as there are more modern and sophisticated methods available.

15.4. Thin layer chromatography

It was developed by Mein hard and Hall it is similar to paper chromatography and has a solid stationary phase and a liquid mobile phase. Here in the case of TLC, slurry of silica or alumina or other compounds are used as a stationary phase and a variety of solvents can be used as mobile phase. The separation is based on the migration of the sample through stationary phase and mobile phase. The components with more affinity elute fast while the less affinity ones elute later. There are two methods, normal phase and reverse phase in normal phase stationary phase is polar and the mobile phase is non-polar in reverse phase stationary phase is non-polar and the mobile phase is the polar. The substances are identified using the retention factor (R_f values) which was calculated using the formula,

$$R_f = \frac{\text{Distance travelled by the solute}}{\text{Distance travelled by the solvent}}$$

This is one of the most used chromatographic methods for the phytochemical analysis.

15.5. Gas Chromatography

Volatile compounds are analysed using gas chromatography. In this method, there is a gas and a liquid phase. The liquid phase is stationary where the gas phase is a mobile phase. These compounds to be analysed are also in the mobile phase with a carrier gas which is usually helium, hydrogen or argon. The chemicals are separated depending on the migration rate into the liquid phase. Higher percentage of the chemical will lead to faster migration in the liquid phase. There are two major types, Gas-solid chromatography – Solid stationary phase and gaseous mobile phase and gas-liquid chromatography - Solid stationary phase and gaseous mobile phase. The gas chromatography is connected with a mass spectrometry which analysis both the compounds present and the quantity of the compounds present .This is widely used in qualitative and quantitative phytochemical analysis. The limitation is as the samples are incinerated the valuable thermolabile secondary metabolites and other novel compounds may get degraded.

15.6. High Performance Liquid Chromatography (HPLC)

HPLC is also known as High- Pressure Liquid Chromatography. This method involves the interaction of liquid solvent in the tightly packed solid column or a liquid column. These acts as the stationary phase while the liquid (solvent) acts as the mobile phase, high pressure enables the compounds to pass to the detector. As HPLC compounds are analysed after vaporisation, thermolabile compounds cannot be analysed with this technique. Separation of compounds based on the difference in affinity of compounds towards stationary phase. Most widely used stationary phase is unmodified silica which allows high efficiency and high permeability. The limitation is similar to that of gas chromatography whereas the samples are incinerated the valuable thermolabile secondary metabolites and other novel compounds may get degraded.

15.7. High Performance Thin Layer Chromatography (HPTLC)

This method is modified form of thin layer chromatography. It is a type of planer chromatography where the separation is done by high performance layers with detection and the sample components are acquisition using an advanced work- station. The reduction of the thickness of the layer will increase the efficiency of the separation and hence HPTLC is more advanced method for qualitative, quantitative and micro-preparative chromatography. Plants

15.8. Optimum Performance Laminar Chromatography (OPLC)

OPLC combines the advantages of TLC and HPLC. The system separates about 10-15 mg samples, with simultaneous processing of up to 4 or 8 samples at a time depending on the model. In OPLC a pump is used to force a liquid mobile phase through a stationary phase, such as silica or a bonded-phase medium.

15.9. References

• CoÅŸkun, Ozlem. (2016). Separation techniques: Chromatography. Northern Clinics of Istanbul. 3. 10.14744/nci.2016.32757.

• Rashid Romana et. al., Extraction Method and Isolation of Phytoconstituents from the Chloroform Extract of Plant Solanum Nigrum L. By Colunm Chromatography. International Journal of Advance Research, Ideas and Innovations in Technology , Volume3, Issue6

• Ozlem Coskun . Separation techniques: Chromatography. North Clin Istanbul 2016;3(2):156–60 doi: 10.14744/nci.2016.32757

• Das M, Dasgupta D. Pseudo-affinity column chromatography based rapid purification procedure for T7 RNA polymerase. Prep Biochem Biotechnol 1998;28:339–48

• Stoddard JM, Nguyen L, Mata-Chavez H, Nguyen K. TLC plates as a convenient platform for solvent-free reactions. Chem Commun (Camb) 2007;12:1240–1.

• Sherman J, Fried B, Dekker M. Handbook of Thin-Layer Chromatography New York, NY; 1991.

• Donald PL, Lampman GM, Kritz GS, Randall G. Engel Introduction to Organic Laboratory Techniques (4th Ed.). ThomsonBrooks/Cole 2006. p. 797–817.

http:/80.251.40.59/veterinary.ankara.edu.tr/f idanci/Ders_Notlari/Biyoteknoloji/Kromatografi.ht ml.

•

https://learning.oreilly.com/library/view/pha rmaceutic analysis /9789332515659/xhtml/chapter034.xhtml

• Regnier FE. High-performance liquid chromatography of biopolimers. Science 1983. p. 245–52.

• .Biradar RS, Rachetti DB (2013). Extraction of some secondary metabolites & thin layer chromatography from different parts of Centella asiatica L. American Journal of Life Sciences. 1(6): 243-247.

• Lukasz C, Monika WH, Review Two-dimensional thinlayer chromatography in the analysis of secondary plant metabolites Journal of Chromatography A, 1216 (2009) 1035–1052

Chapter sixteen

Recent developments in the phytochemical analysis

16.1. Introduction

Plants usually contain hundreds even thousands of compounds that have therapeutic effect or related to a therapeutic effect. The composition and chemical constituents of the plants can vary due to a number of factors like growing condition, environment, species differences and even processing methods which make plant analysis a very complex one. There are so many methods and analytical techniques we use for the analysis of plant compounds. Nowadays modern chromatographic methods are used for the separation and identification of plant compounds. Some of the recent methods for analysis of plant samples are seen in this chapter.

16.2. Recent developments in the phytochemical analysis

One-dimensional (1d) chromatography coupled to MS

Gas chromatography is made advanced by connecting it with a mass spectrometer it increases the speed, accuracy, resolution and sensitivity of the analysis. LC–MS is the most widely used approach for the analysis of medicinal plants. Ultra-high performance liquid chromatography (UHPLC) is a technique which applies the sub-2 μm porous column-packing materials for the analysis of medicinal plants. These are highly applicable for analysis of charged or polar molecules.

Multiple-dimensional chromatography (DMC) combined with MS

The one dimensional chromatography with MS has some disadvantages like co-elution or detection of low concentration compounds and high background. To avoid these limitations, two or more orthogonal separation procedures in a consecutive manner increases accuracy and resolution.

Two-dimensional (2D) LC–MS

The two dimensional liquid chromatography methods can be used for separation and identification of some very complex samples with number of components in different polarities. It has analytical ability and also can be great method to analyse plant samples. It is rapid and robust and can be used in pharma industries for screening potentially bioactive components. It also has applications in application in all aspects of drug-discovery stages, including initial screening of lead compounds, hit-to-lead optimisation, hit-to-lead optimisation, and clinical trials.

Two-dimensional (2D) GC–MS

It can be used for targeted and untargeted identification of compounds, quantitative and qualitative analysis and fingerprinting in plants. 3D-GC has the ability of collecting milligram levels of unknown compound in a rapid manner for the further identification through NMR, MS and Fourier transform ion cyclotron resonance (FT-ICR) can be used.

LC-Solid phase extraction (SPE) - NMR-MS

In this method, the sample was first separated in HPLC and was split into two streams where the first one was trapped by SPE for online or offline NMR analysis. The second stream was directly detected by MS. It features efficient separation and information from both NMR and MS data.

Chip-based separation devices coupled to MS

A microfluidic chip or lab on a chip is a miniature laboratory process and can perform multiple functions from sample preparation to separation and detection. Coupling microfluidic chips with MS can provide high sensitivity, parallel sampling can be done; only small quantity of sample is required and high throughput analysis can be done. This method can reduce the time for preparation of sample and the High throughput analysis can be easily done and can reduce time and cost.

Ambient ionisation MS

Though conventional methods have high advantages there are a number of pre-treatment steps have to be done which makes the process tedious and time consuming. Ambient ionisation MS technique has a set of MS methods, which allows direct analysis of the sample without minimal or no pre-treatment in ambient, open air conditions. Also the cost and time is saved as sample preparation and chromatography is not necessary. The method is non-invasive and capable of spatial imaging it can be coupled with MS including TOF, Orbitraps, and FT-ICR.

Spectral databases and computational analysis

There are number of open access databases are created for compound identification some of them are HMDB, the Platform for RIKEN Metabolomics (PRIMe), MassBank, and Metlin. HMDB is on human metabolites, PRIMe is on plat metabolites, MassBank is dedicated for primary metabolites and natural substances. Metlin deals with data for drug discovery and biomarker discovery. The overall coverage of metabolome is still limited due to complicated methods and authentication standards

16.3. Reference

- Zhu M-Z, Chen G-L, Wu J-L, Li N, Liu Z-H, Guo M-Q. Recent development in mass spectrometry and its hyphenated techniques for the analysis of medicinal plants. Phytochemical

Analysis.   2018;29:365–374.   https://doi.
org/10.1002/pca.2763

•   Wu HF, Guo J, Chen SL, et al. Recent
developments in qualitative and quantitative
analysis of phytochemical constituents and their
metabolites using liquid chromatography-mass
spectrometry.   J   Pharmaceut   Biomed.
2013;72:267-291.

•   Dong X, Wang R, Zhou X, Li P, Yang H.
Current mass spectrometry approaches and
challenges for the bioanalysis of traditional
Chinese medicines. J Chromatogr B.
2016;1026:15-26.

•   Zhu MZ, Dong X, Guo MQ. Phenolic
profiling of Duchesnea indica combining
macroporous resin chromatography (MRC) with
HPLC-ESI-MS/ MS and ESI-IT-MS. Molecules.
2015;20:22463-22475.

•   Zhu MZ, Liu T, Zhang CY, Guo MQ.
Flavonoids of lotus (Nelumbo nucifera) seed
embryos and their antioxidant potential. J Food
Sci. 2017;82:1834-1841.

•   Zhu MZ, Wu W, Jiao LL, Yang PF, Guo
MQ. Analysis of flavonoids in Lotus (Nelumbo

nucifera) leaves and their antioxidant activity using macroporous resin chromatography coupled with LC-MS/MS and antioxidant biochemical assays. Molecules. 2015;20:10553-10565.

• Wolfender JL, Marti G, Thomas A, Bertrand S. Current approaches and challenges for the metabolite profiling of complex natural extracts. J Chromatogr A. 2015;1382:136-164.

• Zhu MZ, Li N, Wang YT, et al. Acid/salt/pH gradient improved resolution and sensitivity in proteomics study using 2D SCX-RP LC-MS. J Proteome Res. 2017;16:3470-3475.

• Iguiniz M, Heinisch S. Two-dimensional liquid chromatography in pharmaceutical analysis. Instrumental aspects, trends and applications. J Pharmaceut Biomed. 2017;145:482-503.

• Sampat A, Lopatka M, Sjerps M, Vivo-Truyols G, Schoenmakers P, van Asten A. Forensic potential of comprehensive two-dimensional gas chromatography. Trac-Trend Anal Chem. 2016;80:345-363.

• Tranchida PQ, Franchina FA, Dugo P, Mondello L. Comprehensive two-dimensional gas chromatography-mass spectrometry: recent evolution and current trends. Mass Spectrom Rev. 2016;35:524-534

• Cao JL, Wei JC, Chen MW, et al. Application of two-dimensional chromatography in the analysis of Chinese herbal medicines. J Chromatogr A. 2014;1371:1-14.

• Liang Z, Li KY, Wang XL, Ke YX, Jin Y, Liang XM. Combination of off- line two-dimensional hydrophilic interaction liquid chromatography for polar fraction and two-dimensional hydrophilic interaction liquid chromatography × reversed-phase liquid chromatography for medium-polar fraction in a traditional Chinese medicine. J Chromatogr A. 2012;1224:61-69.

• Cao JL, Wei JC, Chen MW, et al. Application of two-dimensional chromatography in the analysis of Chinese herbal medicines. J Chromatogr A. 2014;1371:1-14.

• Dong X, Wang R, Zhou X, Li P, Yang H. Current mass spectrometry approaches and challenges for the bioanalysis of traditional Chinese medicines. J Chromatogr B. 2016;1026:15-26

• Sciarrone D, Panto S, Rotondo A, et al. Rapid collection and identification of a novel component from Clausena lansium Skeels leaves by means of three-dimensional preparative gas chromatography and nuclear magnetic resonance/infrared/mass spectrometric analysis. Anal Chim Acta. 2013;785:119-125.

• Panto S, Sciarrone D, Maimone M, et al. Performance evaluation of a versatile multidimensional chromatographic preparative system based on three-dimensional gas chromatography and liquid chromatography-two-dimensional gas chromatography for the collection of volatile constituents. J Chromatogr A. 2015;1417:96-103.

• Sciarrone D, Panto S, Donato P, Mondello L. Improving the productivity of a multidimensional chromatographic preparative system by collecting pure chemicals after each of three chromatographic dimensions. J Chromatogr A. 2016;1475:80-85.

• Seger C, Sturm S, Stuppner H. Mass spectrometry and NMR spectroscopy: modern high-end detectors for high resolution separation techniques – state of the art in natural product

HPLC-MS, HPLC- NMR, and CE-MS hyphenations. Nat Prod Rep. 2013;30:970-987.

• Schlotterbeck G, Ceccarelli SM. LC-SPE-NMR-MS: a total analysis system for bioanalysis. Bioanalysis. 2009;1:549-559.

• Gu WY, Li N, Leung EL, et al. Rapid identification of new minor chemical constituents from Smilacis Glabrae Rhizoma by combined use of UHPLC-Q-TOF-MS, preparative HPLC and UHPLC-SPE-NMR-MS techniques. Phytochem Anal. 2015;26:428-435.

• Haghighi F, Talebpour Z, Sanati-Nezhad A. Through the years with on- a-chip gas chromatography: a review. Lab Chip. 2015;15:2559-2575.

• Wong MY, So PK, Yao ZP. Direct analysis of traditional Chinese medicines by mass spectrometry. J Chromatogr B. 2016;1026:2-14.

• Li XJ, Wang X, Li LN, Bai Y, Liu HW. Direct analysis in real time mass spectrometry: a powerful tool for fast analysis. Mass Spectrom Lett. 2015;6:1-6.

• Yang YY, Deng JW. Analysis of pharmaceutical products and herbal medicines using ambient mass spectrometry. Trac-Trend Anal Chem. 2016;82:68-88.

• Vinaixa M, Schymanski EL, Neumann S, Navarro M, Salek RM, Yanes O. Mass spectral databases for LC/MS- and GC/MS-based metabolomics: state of the field and future prospects. Trac-Trend Anal Chem. 2016;78:23-35.

• Sumner LW, Lei Z, Nikolau BJ, Saito K. Modern plant metabolomics: advanced natural product gene discoveries, improved technologies, and future prospects. Nat Prod Rep. 2015;32:212-229.

Index

www.ingramcontent.com/pod-product-compliance
Lightning Source LLC
Chambersburg PA
CBHW030939180526
45163CB00002B/635